SpringerBriefs in Computer Science

More information about this series at http://www.springer.com/series/10028

Gerardo I. Simari • Cristian Molinaro
Maria Vanina Martinez • Thomas Lukasiewicz
Livia Predoiu

Ontology-Based Data Access Leveraging Subjective Reports

Springer

Gerardo I. Simari
Consejo Nacional de Investigaciones
   Cientificas y Tecnicas
Universidad Nacional del Sur
Bahia Blanca, Argentina

Maria Vanina Martinez
Consejo Nacional de Investigaciones
   Cientificas y Tecnicas
Universidad Nacional del Sur
Bahia Blanca, Argentina

Livia Predoiu
Department of Computer Science
University of Oxford
Oxford, Oxfordshire, UK

Cristian Molinaro
DIMES Department
Università della Calabria
Rende, Italy

Thomas Lukasiewicz
Department of Computer Science
University of Oxford
Oxford, Oxfordshire, UK

ISSN 2191-5768   ISSN 2191-5776 (electronic)
SpringerBriefs in Computer Science
ISBN 978-3-319-65228-3   ISBN 978-3-319-65229-0 (eBook)
DOI 10.1007/978-3-319-65229-0

Library of Congress Control Number: 2017951849

Printed on acid-free paper

This Springer imprint is published by Springer Nature
The registered company is Springer International Publishing AG
The registered company address is: Gewerbestrasse 11, 6330 Cham, Switzerland

Acknowledgments

This work is the result of the collaboration among three research groups in Universidad Nacional del Sur and CONICET (Argentina), Università della Calabria (Italy), and University of Oxford (United Kingdom). In particular, the final efforts to put it together were made possible by CONICET's *Programa de Financiamiento Parcial de Estadías en el Exterior para Investigadores Asistentes*, which allowed Gerardo Simari to visit Università della Calabria for 3 months.

Other sources of funding also came from Universidad Nacional del Sur (UNS), CONICET, Agencia Nacional de Promoción Científica y Tecnológica (Argentina), ONR grant N00014-15-1-2742 (United States Department of the Navy, Office of Naval Research),[1] UK EPSRC grants EP/J008346/1, EP/L012138/1, and EP/M025268/1, and the Alan Turing Institute under EPSRC grant EP/N510129/1.

---

[1] Any opinions, findings, and conclusions or recommendations expressed in this material are those of the authors and do not necessarily reflect the views of the Office of Naval Research.

Contents

Chapter 1
Ontology-Based Data Access with Datalog+/−

Ontologies are logical theories that describe a formal conceptualization in a domain of interest. Conceptualizations are intensional semantic structures that codify implicit knowledge that restricts the structure of a part of the domain. Usually, this specification is expressed explicitly in a (declarative) language. This formalization makes the knowledge available for machine processing, facilitating in this way its interchange. For this reason, in the last years, ontologies have been of increasing interest in many and diverse applications, especially in the context of artificial intelligence (AI), the Semantic Web, and data integration.

A conceptualization is an abstract and simplified view of the world that we intend to represent. Every knowledge base or knowledge-based system is (implicitly or explicitly) associated with a conceptualization. When the domain knowledge is represented in a declarative formalism, the set of objects that can be represented are called the universe of discourse: for these systems, what exists is what can be represented.

In the context of AI or knowledge representation and reasoning (KR&R), we can describe an ontology of a program as a set of representational terms. The definition of entity names in the universe of discourse (classes, relationships, functions, and other objects) are associated on the one hand with descriptions of what the names mean, and on the other hand with formal axioms that constrain the interpretation and the well-formed use of those terms. Formally, we can say that an ontology describes a logical theory.

The formal specification of an ontology comprises several levels:

1. *Meta-level*: this level specifies the set of categories that are modeled.
2. *Intensional level*: specifies the set of conceptual elements (instances of categories) and the rules that describe the structure of the conceptual domain.

The original version of this chapter was revised. An erratum to this chapter can be found at https://doi.org/10.1007/978-3-319-65229-0_5

© Springer International Publishing AG 2017
G.I. Simari et al., *Ontology-Based Data Access Leveraging Subjective Reports*,
SpringerBriefs in Computer Science, DOI 10.1007/978-3-319-65229-0_1

3. *Extensional level*: specifies the set of instances of the conceptual elements described by the intensional level.

The following example shows the components of the three levels in a domain describing restaurants and the food that they serve.

Example 1.1 Consider the task of building a very simple data model for restaurants and food. The different levels in this could be the following:

- *Meta-level*: possible categories to be modeled are: *Restaurant, Restaurant Cuisine, Business, Neighborhood, City, Country, Food*, and *Food Type*.
- *Intensional level*: The following are examples of intensional knowledge—we provide them in natural language, as we have not yet specified an ontological language; we discuss again these elements in the language of specific description logics in Example 1.2:

  - A restaurant is a business.
  - Restaurants serve food.
  - For each place, there exists a neighborhood in which it is located.
  - For each place, there exists a city in which it is located.
  - For each food, there exists an associated type.

- *Extensional level*: the following are examples of instances:

  - *Food*: Pizza Margherita, Chicken Burrito, Soupe à l'oignon, Waldorf Salad.
  - *Food Type*: Sandwich, Salad, Pizza, Pasta, Meat.
  - *Restaurant Cuisine*: Italian, French, Argentine, Asian, Mexican, Vegetarian, Kosher.   ∎

The area of KR&R focuses on methods to provide high level descriptions of the real world, which can be used to build intelligent applications. The focus of KR&R can be divided into two categories: (1) logic-based knowledge-based formalisms, which evolved from the intuition that first-order logic (or logical theories in general) can be used to capture facts about the world in an unambiguous way; and (2) non-logical representations, inspired in rather cognitive notions, for instance, network structures and representations based on rules derived from experiments over human activities (e.g., data-driven).

In the logic-based approach, the representation language is, in general, a variant of first-order logic (FOL), and the task of reasoning implies the verification of logical consequences. On the other hand, the non-logical approaches are based on graphical interfaces, and the knowledge is represented in ad-hoc data structures. Well-known examples of these are semantic networks [34] and frames [33], which characterize the knowledge and reasoning by means of cognitive structures in the form of networks. From the practical point of view, systems based on networks are more attractive, however, their lack of a semantic characterization is an important issue: different systems, whose components look much alike (for instance, having similar or identical names for relations), behave in a very different way. Nevertheless, it is possible to provide them with a FOL semantics (at least to a central set of their characteristics), this is where description logics (DLs) are born, from the combination of both approaches.

1.1 Description Logics

The development of description logics (DLs) began under the name of *termino-logical systems*, establishing the basic terminology adopted in the modeling of a domain. Afterwards, the development focused on the set of constructors that would build the concepts admitted in the language. More recently, the attention in the area changed to the advancement of the properties of the underlying logical systems, and therefore the term description logics was coined.

DLs are a family of knowledge representation formalisms based on logic [3, 4]. They describe the domain of interest in terms of concepts (classes), roles (relations), and individuals. Their formal semantics is based on model theory, as they correspond to decidable fragments of FOL; but they are also related to propositional modal logics and dynamic logics. As in any logical theory, we can make inferences about its contents. From the computational point of view, there exist sound and complete procedures for specific reasoning tasks, such as logical inference, query answering, and inconsistency-tolerant inference. For some DLs, specifically for logical inference and query answering, there are implemented systems with a high degree of optimization, which allows to solve Big Data problems [1] and problems within the Semantic Web [10].

The basic syntactic step in DLs is provided by two different symbol alphabets: one denotes atomic concepts by means of unary predicate symbols, and the other denotes atomic roles, designated by binary predicate symbols, which are used to express relationships between concepts. In the following section, we present Datalog+/−, an ontology language based on logical rules that generalizes several DLs and allows for predicate symbols with arbitrary arity. Terms are then built from the basic symbols using several constructors. For instance, the intersection of concepts C and D, which is denoted $C \sqcap D$, restricts the set of individuals to those that belong to both C and D. Note that, in the syntax of DLs, concept expressions are variable-free. In fact, a concept expression denotes the set of all individuals satisfying the properties specified by the expression. This means that the expression $C \sqcap D$ is equivalent to the FOL expression $C(X) \wedge D(X)$, where X ranges over all individuals in the interpretation domain, and $D(X)$ is true for all individuals that belong to the concept D.

The set of constructs for establishing the relationships between concepts is what characterizes the different DLs. The constructs are the features that determine the expressive power of DLs; Table 1.1 shows some of the most common ones.

The formal semantics of DLs is given in terms of FOL interpretations. An interpretation $I = (\Delta^{\mathcal{I}}, \cdot^{\mathcal{I}})$ consists of:

- A nonempty set $\Delta^{\mathcal{I}}$, with domain $\mathcal{I}$.
- An interpretation function $\cdot^{\mathcal{I}}$ that maps: (1) each individual a to an element $a^{\mathcal{I}}$ in $\Delta^{\mathcal{I}}$, (2) each atomic concept A to a subset $A^{\mathcal{I}}$ of $\Delta^{\mathcal{I}}$, (3) each atomic role P to a subset $P^{\mathcal{I}}$ of $\Delta^{\mathcal{I}} \times \Delta^{\mathcal{I}}$.
- The interpretation function extends to complex concepts and roles according to the syntactic structure.

Table 1.1 Syntax and semantics of the most common DLs constructs

Construct	Syntax	Semantics
Atomic concept	A	$A^{\mathcal{I}} \subseteq \Delta^{\mathcal{I}}$
Atomic role	P	$P^{\mathcal{I}} \subseteq \Delta^{\mathcal{I}} \times \Delta^{\mathcal{I}}$
Atomic negation	$\neg A$	$\Delta^{\mathcal{I}} - A^{\mathcal{I}}$
Concept inclusion	$C \sqsubseteq D$	$\forall a.a \in C^{\mathcal{I}} \to a \in D^{\mathcal{I}}$
Conjunction	$C \sqcap D$	$C^{\mathcal{I}} \cap D^{\mathcal{I}}$
(Unqual.) existential restriction	$\exists R$	$\{a \mid \exists b.(a,b) \in R^{\mathcal{I}}\}$
Value restriction	$\forall R.C$	$\{a \mid \forall b.(a,b) \in R^{\mathcal{I}} \to b \in C^{\mathcal{I}}\}$
Bottom	$\bot$	$\emptyset$
Disjunction	$C \sqcup D$	$C^{\mathcal{I}} \cup D^{\mathcal{I}}$
Top	$\top$	$\Delta^{\mathcal{I}}$
Qualified existential restriction	$\exists R.C$	$\{a \mid \exists b.(a,b) \in R^{\mathcal{I}} \wedge b \in C^{\mathcal{I}}\}$
Full negation	$\neg C$	$\Delta^{\mathcal{I}} - C^{\mathcal{I}}$
Inverse role	R^{-}	$\{(a,b) \mid (b,a) \in R^{\mathcal{I}}\}$

The main differences w.r.t. conventional (relational) databases are that the interpretation domain is arbitrary and can be infinite, and furthermore, there is an underlying open-world assumption (i.e., the knowledge is not complete). The third column in Table 1.1 shows the semantics for a set of common constructs.

A DL is characterized by an ABox (usually denoted with $\mathcal{A}$), a mechanism to specify properties about objects (assertional axioms), and a TBox (usually denoted with $\mathcal{T}$), a mechanism to specify the knowledge about relationships among concepts and roles (terminological axioms). Concept inclusions are the most basic assertions within a TBox, denoted with $C \sqsubseteq D$, which states that the concept D is more general than the concept C, which means that every individual that is an instance of the concept C is always also an instance of the concept D. Another commonly used assertion is functionality on roles, denoted $funct(R)$, establishing that the relation that the role R specifies is a function. Figure 1.1 shows other DL assertions that can be used to express relationships between complex concepts in a TBox $\mathcal{T}$. The ABox $\mathcal{A}$ and the TBox $\mathcal{T}$ together form a knowledge base denoted with $\mathcal{K} = (\mathcal{A}, \mathcal{T})$. Furthermore, a DL provides a set of inference services that allow us to reason about the contents of the knowledge base $\mathcal{K}$.

Consider the following example, where the abstract conceptualization described in Example 1.1 is specified by means of a DL ABox and TBox.

Example 1.2 From Example 1.1, we can define the following concepts and roles:

- Concepts: *Food, Restaurant, Business, Cuisine, FoodType, Neighborhood, City,* and *Country*.
- Roles: *Type* (relates food with its corresponding food type), *restaurantCuisine, serves* (associates restaurants with the food they serve in their menu), *LocatedIn* (associates businesses to neighborhoods or cities, and neighborhoods (resp., cities) to cities (resp., countries)).

Description Logic Assertion	Datalog+/– Rule
CONCEPT INCLUSION: $Restaurant \sqsubseteq Business$	$restaurant(X) \rightarrow business(X)$
CONCEPT PRODUCT: $Food \times Food \sqsubseteq TwoCourseMeal$	$food(X), food(Y) \rightarrow twoCourseMeal(X,Y)$
INVERSE ROLE INCLUSION: $InPromotionIn^{-} \sqsubseteq Serves$	$inPromotionIn(F,R) \rightarrow serves(R,F)$
ROLE TRANSITIVITY: $trans(LocatedIn)$	$locatedIn(X,Y), locatedIn(Y,Z) \rightarrow locatedIn(X,Z)$
PARTICIPATION: $Restaurant \sqsubseteq \exists Serves.Food$	$restaurant(R) \rightarrow \exists F \; serves(R,F) \wedge food(F)$
DISJOINTNESS: $City \sqcap Country \sqsubseteq \bot$	$city(X), country(X) \rightarrow \bot$
FUNCTIONALITY: $funct(LocatedIn)$	$locatedIn(X,Y), locatedIn(X,Z) \rightarrow Y = Z$

Fig. 1.1 Translation of several different types of description logic axioms into Datalog+/–

$$\mathcal{A} = \{ \; Food(bifeDeChorizo), Food(soupAlOignon), FoodType(meat),$$
$$City(buenosAires), RestaurantCuisine(laCabrera, argentine),$$
$$Serves(laCabrera, bifeDeChorizo), Restaurant(laCabrera),$$
$$Restaurant(laTartine) \}.$$

$$\mathcal{T} = \{ \; Restaurant(R) \sqsubseteq Business(R),$$
$$Serves^{-} \sqsubseteq Food,$$
$$Restaurant \sqsubseteq \exists Serves,$$
$$RestaurantCuisine^{-} \sqsubseteq Cuisine,$$
$$Restaurant \sqsubseteq \exists RestaurantCuisine,$$
$$LocatedIn^{-} \sqsubseteq City,$$
$$Business(B) \sqsubseteq \exists LocatedIn,$$
$$func(locatedIn),$$
$$City \sqcap Country \sqsubseteq \bot \}.$$

The ABox $\mathcal{A}$ contains the set of instances of concepts and roles in this domain; for instance, we have that *bifeDeChorizo* and *soupAlOignon* are instances of *food*, and *laCabrera* is an instance of *restaurant*. Regarding roles, we have that *laCabrera* serves *bifeDeChorizo* in its menu.

In the TBox $\mathcal{T}$, the terminological axioms state that every *Restaurant* is also a *Business*, every *Restaurant* serves at least one *Food*, the concepts *City* and *Country* are disjoint (no city is a also country, and vice versa), and every *City* is located in exactly one *Country*, among others. ∎

1.2   Ontology-Based Data Access (OBDA)

Traditionally, the ad hoc manner in which data stores are accessed involves dedicating an application to extract the data from the data source. In the presence of processes that need to access several, potentially heterogeneous sources, this ad hoc method is generalized as shown in Fig. 1.2a, where several (potentially redundant) applications access data sources individually. Since data are in general heterogeneous, distributed, redundant, inconsistent, and incoherent, it then becomes difficult to reason about them as a whole. Furthermore, each additional application and/or data source will involve setting up new—potentially complex—data access modules.

Alternatively, the usage of KR principles and techniques based on logic, can provide a conceptual, high-level representation of the domain of interest in terms of an ontology; this is known as ontology-based data access (OBDA). OBDA allows us to answer queries from (incomplete) data sources in the presence of domain specific knowledge provided by an ontology. That is, the query is answered by taking into account the information described in an ontology from both the axiomatic and the terminological parts. Following this approach, there is no need to migrate the data, which is left at the sources; rather a mapping is defined between an ontology and the data sources. The basic idea is to use a rich semantic formalism to hide where and how the data is stored, while presenting to the user a conceptual view of that data. In this way, any information access to the data is specified in terms of the ontology, and the inference services of the OBDA system is used to translate the requests into queries to the data sources. Figure 1.2b shows an overview of how such systems work.

In this work, we provide an OBDA approach to reasoning about data contained in a collection of subjective reports regarding application-dependant objects. The logical abstraction in terms of an ontology enriched with preferences allows us to build customized answers to queries posted by a user that interpret these reports in terms of the user's preferences. In particular, our proposal builds on top of Datalog+/− [16], a family of rule-based ontological languages that cover several important DLs, bridging the gap in expressive power between database query languages and DLs as ontology languages, and extending the well-known Datalog language in order to embed DLs.

1.3   The Datalog+/− Family of Ontology Languages

We now present the basics of Datalog+/− [16]—relational databases, (Boolean) conjunctive queries, tuple- and equality-generating dependencies and negative constraints, the chase, and ontologies. Part of the material presented in this section is based on [36].

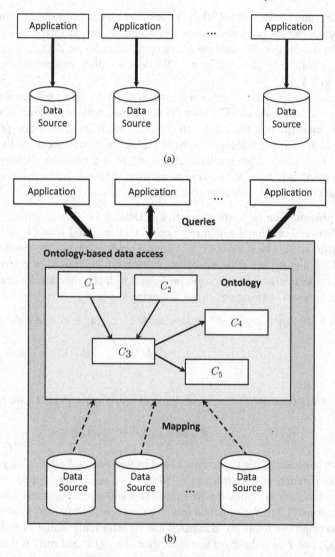

Fig. 1.2 (**a**) Traditional data access: each application has a separate access set up. (**b**) OBDA: Different data sources are accessed through a domain conceptualization—an ontology

1.3.1   Preliminary Concepts and Notations

Let us consider (1) an infinite universe of *(data) constants* Δ, which constitute the "normal" domain of a database), (2) an infinite set of *(labelled) nulls* Δ_N (used as "fresh" Skolem terms, which are placeholders for unknown values, and can thus be seen as variables), and (3) an infinite set of variables $\mathcal{V}$ (used in queries, dependencies, and constraints). Different constants represent different values (this is generally known as the *unique name assumption*), while different nulls may represent the same value. We assume a lexicographic order on $\Delta \cup \Delta_N$, with every symbol in Δ_N following all symbols in Δ. We denote with $\mathbf{X}$ sequences of variables $X_1, \ldots, X_k$ with $k \geq 0$.

We will assume a *relational schema* $\mathcal{R}$, which is a finite set of *predicate symbols* (or simply *predicates*). As usual, a *term t* is a constant, null, or variable. An *atomic formula* (or *atom*) a has the form $p(t_1, \ldots, t_n)$, where p is an n-ary predicate, and $t_1, \ldots, t_n$ are terms. A term or atom is *ground* if it contains no nulls and no variables. An *instance I* for a relational schema $\mathcal{R}$ is a (possibly infinite) set of atoms with predicates from $\mathcal{R}$ and arguments from $\Delta \cup \Delta_N$. A *database* is a finite instance that contains only constants (i.e., its arguments are from Δ).

Homomorphisms Central to the semantics of Datalog+/− is the notion of *homomorphism* between relational structures. Let $A = \langle X, \sigma^A \rangle$ and $B = \langle Y, \sigma^B \rangle$ be two relational structures, where $dom(A) = X$ and $dom(B) = Y$ are the domains of A and B, and σ^A and σ^B are their signatures (which are composed of relations and functions), respectively. A *homomorphism* from A to B is a function $h : dom(A) \to dom(B)$ that "preserves structure" in the following sense:

- For each n-ary function $f^A \in \sigma^A$ and elements $x_1, \ldots, x_n \in dom(A)$, we have:

$$h(f^A(x_1, \ldots, x_n)) = f^B(h(x_1), \ldots, h(x_n)),$$

  and
- For each n-ary relation $R^A \in \sigma^A$ and elements $x_1, \ldots, x_n \in dom(A)$, we have:

$$\text{if } (x_1, \ldots, x_n) \in R^A, \text{ then } (h(x_1), \ldots, h(x_n)) \in R^B.$$

In the above statements, the superscripts used in function and relation symbols are simply a clarification of the structure in which they are being applied. Since we do not have function symbols, the first condition will not be necessary here (it is satisfied vacuously). The fundamental result linking homomorphisms to conjunctive query answering over relational databases can be informally stated as follows: let Q be a BCQ, and J be a database instance; then, $J \models Q$ if and only if there exists a homomorphism from the *canonical database instance I^Q* (essentially, an instance built using the predicates and variables from Q) to J [2, 18].

For the purposes of Datalog+/−, we need to extend the concept of homomorphism to contemplate nulls. We then define homomorphisms from a set of atoms A_1 to a set of atoms A_2 as mappings $h: \Delta \cup \Delta_N \cup \mathcal{V} \to \Delta \cup \Delta_N \cup \mathcal{V}$ such that:

1. $c \in \Delta$ implies $h(c) = c$,
2. $c \in \Delta_N$ implies $h(c) \in \Delta \cup \Delta_N$,
3. $r(t_1, \ldots, t_n) \in A_1$ implies $h(r(t_1, \ldots, t_n)) = r(h(t_1), \ldots, h(t_n)) \in A_2$.

Similarly, one can extend h to a conjunction of atoms. Conjunctions of atoms are often identified with the *sets* of their atoms.

1.3.2   Syntax and Semantics of Datalog+/−

Given a relational schema $\mathcal{R}$, a Datalog+/− program consists of a finite set of tuple-generating dependencies (TGDs), negative constraints (NCs), and equality-generating dependencies (EGDs).

A *tuple-generating dependency* (TGD) σ is a first-order (FO) rule that allows existentially quantified conjunctions of atoms in rule heads:

$$\sigma \; : \; \forall \mathbf{X} \forall \mathbf{Y} \; \underbrace{\Phi(\mathbf{X}, \mathbf{Y})}_{\text{body}(\sigma)} \to \underbrace{\exists \mathbf{Z} \, \Psi(\mathbf{X}, \mathbf{Z})}_{\text{head}(\sigma)} \text{ with } \mathbf{X}, \mathbf{Y}, \mathbf{Z} \subseteq \mathcal{V},$$

where $\Phi(\mathbf{X}, \mathbf{Y})$ and $\Psi(\mathbf{X}, \mathbf{Z})$ are conjunctions of atoms. Note that TGDs with multiple single atoms in the head can be converted into sets of TGDs with only single atom in the head [17]. Therefore, from now on, we assume that all sets of TGDs have only a single atom in their head. An instance I for $\mathcal{R}$ *satisfies* σ, denoted $I \models \sigma$, if whenever there exists a homomorphism h that maps the atoms of $\Phi(\mathbf{X}, \mathbf{Y})$ to atoms of I, there exists an extension h' of h that maps $\Psi(\mathbf{X}, \mathbf{Z})$ to atoms of I.

A *negative constraint* (NC) ν is a first-order rule that allows to express negation:

$$\nu \; : \; \underbrace{\forall \mathbf{X} \, \Phi(\mathbf{X})}_{\text{body}(\nu)} \to \bot \text{ with } \mathbf{X} \subseteq \mathcal{V},$$

where $\Phi(\mathbf{X})$ a conjunction of atoms. An instance I for $\mathcal{R}$ *satisfies* ν, denoted $I \models \nu$, if for each homomorphism h, $h(\Phi(\mathbf{X}, \mathbf{Y})) \not\subseteq I$ holds.

An *equality-generating dependency* (EGD) μ is a first-order rule of the form:

$$\mu \; : \; \underbrace{\forall \mathbf{X} \, \Phi(\mathbf{X})}_{\text{body}(\mu)} \to X_i = X_j \text{ with } X_i, X_j \in \mathbf{X} \subseteq \mathcal{V},$$

where $\Phi(\mathbf{X})$ is conjunction of atoms. An instance I for $\mathcal{R}$ *satisfies* μ, denoted $I \models \mu$, if whenever there is a homomorphism h such that $h(\Phi(\mathbf{X}, \mathbf{Y})) \subseteq I$, it holds that $h(X_i) = h(X_j)$.

In the following, we will sometimes omit the universal quantification in front of TGDs, NCs, and EGDs, and assume that all variables appearing in the body are universally quantified.

A *Datalog+/− program* Σ is a finite set $\Sigma_T \cup \Sigma_{NC} \cup \Sigma_E$ of TGDs, NCs, and EGDs. The schema of Σ, denoted $\mathcal{R}(\Sigma)$, is the set of predicates occurring in Σ. A *Datalog+/− ontology* $O = (D, \Sigma)$ consists of a finite database D and a Datalog+/− program Σ. The following example illustrates a simple Datalog+/− ontology, used in the sequel as a running example.

Example 1.3 Consider the ontology $KB = (D, \Sigma)$, where D and $\Sigma = \Sigma_T \cup \Sigma_E$ are defined as follows:

$$\Sigma_T = \{\ r_1 : restaurant(R) \rightarrow business(R),$$
$$r_2 : restaurant(R) \rightarrow \exists F\ food(F) \wedge serves(R, F),$$
$$r_3 : restaurant(R) \rightarrow \exists C\ cuisine(C) \wedge restaurantCuisine(R, C),$$
$$r_4 : business(B) \rightarrow \exists C\ city(C) \wedge locatedIn(B, C),$$
$$r_5 : city(C) \rightarrow \exists D\ country(D) \wedge locatedIn(C, D)\},$$

$$\Sigma_E = \{\ locatedIn(X, Y), locatedIn(X, Z) \rightarrow Y = Z\},$$

$$D = \{\ food(bifeDeChorizo),\quad food(soupAlOignon),$$
$$foodType(meat),\quad foodType(soup),$$
$$cuisine(argentine),\quad cuisine(french),$$
$$restaurant(laCabrera),\quad restaurant(laTartine),$$
$$city(buenosAires),\quad city(paris),$$
$$country(argentina),\quad country(france),$$
$$locatedIn(laCabrera, buenosAires),$$
$$serves(laCabrera, bifeDeChorizo),\quad serves(laTartine, soupeAlOignon)\}.$$

As mentioned in Examples 1.1 and 1.2, this ontology models a very simple knowledge base for restaurants—it could be used, for instance, as the underlying model in an online recommendation and reviewing system (e.g., in the style of TripAdvisor or Yelp). ∎

The conjunction of the first-order sentences associated with the rules of a Datalog$+/-$ program Σ is denoted Σ_P. A *model* of Σ is an instance for $\mathcal{R}(\Sigma)$ that satisfies Σ_p. For a database D for $\mathcal{R}$, and a set of TGDs Σ on $\mathcal{R}$, the set of *models* of D and Σ, denoted $mods(D, \Sigma)$, is the set of all (possibly infinite) instances I such that:

1. $D \subseteq I$, and
2. every $\sigma \in \Sigma$ is satisfied in I (i.e., $I \models \Sigma$).

The ontology is *consistent* if the set $mods(D, \Sigma)$ is not empty.

The semantics of Σ on an input database D, denoted $\Sigma(D)$, is a model I of D and Σ such that for every model I' of D and Σ there exists a homomorphism h such that $h(I) \subseteq I'$; such an instance is called *universal model* of Σ w.r.t. D. Intuitively, a universal model contains no more and no less information than what the given program requires.

In general, there exists more than one universal model of Σ w.r.t. D, but the universal models are (by definition) the same up to homomorphic equivalence, i.e., for each pair of universal models M_1 and M_2, there exist homomorphisms h_1 and h_2 such that $h_1(M_1) \subseteq M_2$ and $h_2(M_2) \subseteq M_1$. Thus, $\Sigma(D)$ is unique up to homomorphic equivalence.

1.3.2.1   Conjunctive Query Answering

We now introduce conjunctive query answering for Datalog+/−. A *conjunctive query* (CQ) over $\mathcal{R}$ has the form:

$$\underbrace{q(\mathbf{X})}_{\text{head}} = \underbrace{\exists \mathbf{Y}\, \Phi(\mathbf{X}, \mathbf{Y})}_{\text{body}},$$

where $\Phi(\mathbf{X}, \mathbf{Y})$ is a conjunction of atoms (possibly equalities, but not inequalities) with the variables $\mathbf{X}$ and $\mathbf{Y}$, and possibly constants, but without nulls, and q is a predicate not occurring in $\mathcal{R}$. A *Boolean CQ* (BCQ) over $\mathcal{R}$ is a CQ of the form $q()$, often written as the set of all its atoms, without quantifiers.

The set of *answers* to a CQ $q(\mathbf{X}) = \exists \mathbf{Y}\, \Phi(\mathbf{X}, \mathbf{Y})$ over an instance I, denoted $q(I)$, is the set of all tuples t over Δ, for which there exists a homomorphism $h \colon \mathbf{X} \cup \mathbf{Y} \to \Delta \cup \Delta_N$ such that $h(\Phi(\mathbf{X}, \mathbf{Y})) \subseteq I$ and $h(\mathbf{X}) = t$. The *answer* to a BCQ $q()$ over a database instance I is *Yes*, denoted $D \models q$, if $q(I) \neq \emptyset$.

Formally, *query answering* under TGDs, i.e., the evaluation of CQs and BCQs on databases under a set of TGDs, is defined as follows. The set of *answers* to a CQ q over a database D and a set of TGDs Σ, denoted $ans(q, D, \Sigma)$, is the set of all tuples t such that $t \in q(I)$ for all $I \in mods(D, \Sigma)$. The *answer* to a BCQ q over D and Σ is *Yes*, denoted $D \cup \Sigma \models q$, if $ans(q, D, \Sigma) \neq \emptyset$. Note that for query answering, homomorphically equivalent instances are indistinguishable, i.e., given two instances I and I' that are the same up to homomorphic equivalence, $q(I)$ and $q(I')$ coincide. Therefore, queries can be evaluated on any universal model.

The decision problem of *CQ answering* is defined as follows: given a database D, a set Σ of TGDs, a CQ q, and a tuple of constants t, decide whether $t \in ans(q, D, \Sigma)$.

For query answering of BCQs in Datalog+/− with TGDs, adding negative constraints is computationally easy, as for each constraint $\forall \mathbf{X} \Phi(\mathbf{X}) \to \bot$ one only has to check that the BCQ $\exists \mathbf{X}\, \Phi(\mathbf{X})$ evaluates to false in D under Σ; if one of these checks fails, then the answer to the original BCQ q is true, otherwise the constraints can simply be ignored when answering the BCQ q.

Adding EGDs over databases with TGDs along with negative constraints does not increase the complexity of BCQ query answering as long as they are *non-conflicting* [16]. Intuitively, this ensures that, if the chase (described next) fails (due to strong violations of EGDs), then it already fails on the database, and if it does not fail, then whenever "new" atoms are created in the chase by the application of the EGD chase rule, atoms that are logically equivalent to the new ones are guaranteed to be generated also in the absence of the EGDs, guaranteeing that EGDs do not influence the chase with respect to query answering. Therefore, from now on, we assume that all the fragments of Datalog+/− have non-conflicting rules.

There are two ways of processing rules to answer queries: *forward chaining* (the *chase*) and *backward chaining*, which uses the rules to rewrite the query in different ways with the aim of producing a query that directly maps to the facts. The key operation is the unification between part of a current goal (a conjunctive query or a fact) and a rule head. Here, we will only cover the chase procedure, which is described next.

1.3.2.2 The TGD Chase

Query answering under general TGDs is undecidable [9] and the chase is used as a procedure to do query answering for Datalog+/−. Given a program Σ with only TGDs (see [16] for further details and for an extended chase with also EGDs), $\Sigma(D)$ can be defined as the least fixpoint of a monotonic operator (modulo homomorphic equivalence). This can be achieved by exploiting the *chase* procedure, originally introduced for checking implication of dependencies, and for checking query containment [23]. Roughly speaking, it executes the rules of Σ starting from D in a forward chaining manner by inferring new atoms, and inventing new null values whenever an existential quantifier needs to be satisfied. By "chase", we refer both to the procedure and to its output.

Let D be a database and σ a TGD of the form $\Phi(\mathbf{X}, \mathbf{Y}) \to \exists \mathbf{Z} \, \Psi(\mathbf{X}, \mathbf{Z})$. Then, σ is *applicable* to D if there exists a homomorphism h that maps the atoms of $\Phi(\mathbf{X}, \mathbf{Y})$ to atoms of D. Let σ be applicable to D, and h_1 be a homomorphism that extends h as follows: for each $Z_j \in \mathbf{Z}$, $h_1(Z_j) = z_j$, where z_j is a "fresh" null, i.e., $z_j \in \Delta_N$, z_j does not occur in D, and z_j lexicographically follows all other nulls already introduced. The *application of* σ on D adds to D the atom $h_1(\Psi(\mathbf{X}, \mathbf{Z}))$ if not already in D. The chase rule described above is also called *oblivious*.

The chase algorithm for a database D and a set of TGDs Σ consists of an exhaustive application of the TGD chase rule in a breadth-first (level-saturating) fashion, which outputs a (possibly infinite) chase for D and Σ.

Formally, the *chase of level up to* 0 of D relative to Σ, denoted $chase^0(D, \Sigma)$, is defined as D, assigning to every atom in D the *(derivation) level* 0. For every $k \geq 1$, the *chase of level up to* k of D relative to Σ, denoted $chase^k(D, \Sigma)$, is constructed as follows: let $I_1, \ldots, I_n$ be all possible images of bodies of TGDs in Σ relative to some homomorphism such that (1) $I_1, \ldots, I_n \subseteq chase^{k-1}(D, \Sigma)$ and (2) the highest level of an atom in every I_i is $k - 1$; then, perform every corresponding TGD application on $chase^{k-1}(D, \Sigma)$, choosing the applied TGDs and homomorphisms in a (fixed) linear and lexicographic order, respectively, and assigning to every new atom the *(derivation) level* k. The *chase* of D relative to Σ, denoted $chase(D, \Sigma)$, is defined as the limit of $chase^k(D, \Sigma)$ for $k \to \infty$.

The (possibly infinite) chase of a database D relative to a set of TGDs Σ, denoted $chase(D, \Sigma)$, is a *universal model*, i.e., there is a homomorphism from $chase(D, \Sigma)$ onto every $B \in mods(D, \Sigma)$ [16] (cf. Fig. 1.3). Thus, BCQs q over D and Σ can be evaluated on the chase for D and Σ, i.e., $D \cup \Sigma \models q$ is equivalent to $chase(D, \Sigma) \models q$. We will assume that the nulls introduced in the chase are named via Skolemization—this has the advantage of making the chase unique; Δ_N is therefore the set of all possible nulls that may be introduced in the chase.

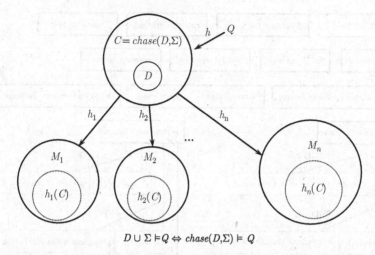

$$D \cup \Sigma \vDash Q \Leftrightarrow chase(D,\Sigma) \vDash Q$$

Fig. 1.3 The chase procedure yields a data structure—also commonly referred to as the chase—that allows to answer queries to a Datalog+/− ontology; it is a universal model, which means that it homomorphically maps to all possible models of the ontology

Example 1.4 Figure 1.4 shows the application of the chase procedure over the Datalog+/− ontology from Example 1.3. As an example, the TGD r_1 is applicable in D, since there is a mapping from atoms *restaurant(laCabrera)* and *restaurant(laTartine)* to the body of the rule. The application of r_1 generates atoms *business(laCabrera)* and *business(laTartine)*.

Consider the following BCQ:

$$q() = \exists X \; restaurant(laTartine) \wedge locatedIn(laTartine, X),$$

asking if there exists a location for restaurant *laTartine*. The answer is *Yes*; in the chase, we can see that after applying TGDs r_1 and r_4, we obtain the atom *locatedIn(laTartine, z_6)*, where z_6 is a null—we would also obtain the same answer, if we ask for restaurant *laCabrera*, because atom *locatedIn(laCabrera, z_5)* is also produced.

Now, consider the CQ:

$$q'(X, Y) = restaurant(X) \wedge locatedIn(X, Y);$$

in this case, we only obtain one answer, namely *(laCabrera, buenosAires)*, since null z_5 eventually maps to *buenosAires*. In the case of *laTartine*, we do not obtain an answer corresponding to the city where it is located: we know there exists a city, but we cannot say which one it is; in every model of *KB*, z_6 takes a different value from the domain. This can be seen in the chase (a universal model of *KB*), since z_6 does not unify with a constant after "chasing" D with Σ.   ∎

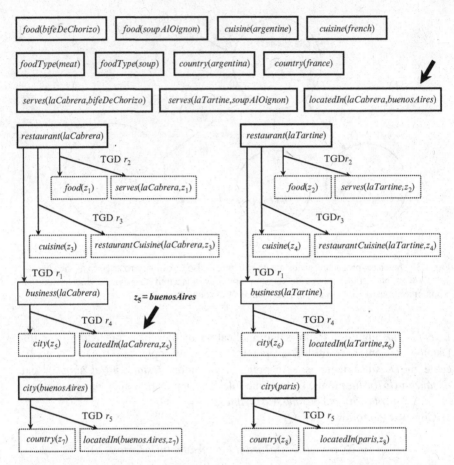

Fig. 1.4 The chase for the ontology in Example 1.3. The atoms in *boxes with thicker border* are part of the database, while those with *dotted lines* correspond to atoms with null values (denoted with z_i). The *arrows* point to the mapping of z_5 to the constant "*buenosAires*" during the chase procedure

1.3.2.3 Computational Complexity

The following complexity measures, partially proposed by Vardi [37], are commonly adopted in the literature:

- The *combined complexity* of CQ answering is calculated by considering all the components—the database, the set of dependencies, and the query—as part of the input.
- The *bounded-arity combined complexity* (or *ba-combined complexity*) is calculated by assuming that the arity of the underlying schema is bounded by an integer constant. In the context of DLs, the combined complexity is equivalent to the *ba*-combined complexity, since the arity of the underlying schema is at most

two. In practical applications, the schema is usually small, and it can safely be assumed to be fixed.

- The *fixed-program combined complexity* (or *fp-combined complexity*) is calculated by considering the set of dependencies to be fixed.
- The *data complexity* is calculated by taking only the database as input.

Some key facts about complexity and decidability of query answering with TGDs: (1) under general TGDs, the problem is undecidable [9], even when the query and set of dependencies are fixed [17]; (2) the two problems of CQ and BCQ evaluation under TGDs are LOGSPACE-equivalent [11]; and (3) the query output tuple (QOT) problem (as a decision version of CQ evaluation) and BCQ evaluation are AC_0-reducible to each other. Given the last two points, we focus only on BCQ evaluation, and any complexity results carry over to the other problems.

1.3.3   Datalog+/− Fragments: Towards Decidability and Tractability

We will now briefly discuss different restrictions that are designed to ensure decidability and tractability of conjunctive query answering with TGDs. While the addition of existential quantifiers in the heads of rules accounts for the "+" in Datalog+/−, these restrictions account for the "−".

Generally, restrictions can be classified into either *abstract* (semantic) or *concrete* (syntactic) properties. Three abstract properties are considered in [7]: (1) the chase is finite, yielding *finite expansion sets* (fes); (2) the chase may not halt but the facts generated have a tree-like structure, yielding *bounded tree-width sets* (bts); and (3) a backward chaining mechanism halts in finite time, yielding *finite unification sets* (fus). Other abstract fragments are: (4) *parsimonious sets* (ps) [29], where the main property for this class is that the chase can be precociously terminated, and (5) *weakly-chase-sticky* TGDs [32] that considers information about the finiteness of predicate positions (positions are infinite if there is an instance D for which an unlimited number of different values appear in that position during the chase).

The main syntactic conditions on TGDs that guarantee the decidability of CQ answering are: (1) *guardedness* [11, 12], (2) *stickiness* [15], and (3) *acyclicity*—each of these classes has a "weak" counterpart: *weak guardedness* [17], *weak stickiness* [15], and *weak acyclicity* [21, 22].

Classes of TGDs Based on Guardedness The following classes are based on this syntactic property:

- *Guarded TGDs* (G) [16]: A TGD is *guarded*, if it has a body atom that contains all the variables appearing in the body—such atoms are called "guards". The class of guarded TGDs is defined as the family of all possible sets of guarded TGDs.

- *Linear TGDs* (L) [16]: An important subclass of guarded TGDs are the so-called *linear* TGDs, which have a single body atom—note that this is automatically a guard, and the restriction allows for self-joins and variable repetition both in the head and the body of a TGD.
- *Weakly guarded TGDs* (WG) [17]: Informally, weakly guarded TGDs extend guarded TGDs by requiring only "harmful" body variables to appear in the guard; guards only need to cover all variables occurring in *affected positions*, where affected positions in predicates are those that may contain fresh labelled nulls generated during the construction of the chase. We have that L ⊂ G ⊂ WG.

  Unfortunately, the property cannot be checked individually for each TGD (as it can be done for L and G). In order to detect the affected positions, one needs to take all the TGDs into account.
- *Frontier-Guarded TGDs* (YG) [6]: Frontier guardedness relaxes the guardedness conditions by requiring the guard atom to contain only the *frontier* of the TGD, which is the set of all variables that appear in both the body and the head of a TGD.
- *Weakly Frontier-Guarded TGDs* (WYG) [6]: A set of TGDs is *weakly frontier-guarded* if, for each TGD, there is an atom in its body that contains all affected variables from the frontier of the TGD. Note that YG ⊂ WYG and WG ⊂ WYG.
- *Disconnected TGDs* (D) [5]: This class contains TGDs that have an empty frontier, but the body and the head may share constants. Note that D ⊂ YG.
- *Joint frontier guarded TGDs* (JYG) [27]: This is a more general class of weakly-guarded and weakly-frontier-guarded, with WYG ⊂ JYG.
- *Frontier-one TGDs* (Y1) [7]: TGDs must have a frontier (as defined above) of size one.
- *Classes characterized by glut variables* [27, 28]: Variables can be divided into *glut* and *non-glut*; the former may take infinitely many values, while the latter can only represent finitely many values. Two classes arise from this notion, namely *glut-guardeness* (gG) and *glut-frontier-guarded* (gYG).

Classes of TGDs Based on Acyclicity The *position dependency graph* of a set of TGDs Σ has nodes representing positions in predicates—node (p, i) represents position i in a predicate p, where p is an n-ary predicate and $1 \leq i \leq n$. For each TGD σ of the form $\Phi(\mathbf{X}, \mathbf{Y}) \to \exists \mathbf{Z} \, \Psi(\mathbf{X}, \mathbf{Z})$ and each variable X in its body occurring in position (p, i), edges with origin (p, i) are built as follows: if $X \in \mathbf{X}$, there is an edge from (p, i) to each position of X in the head of σ. Furthermore, for each $Z \in \mathbf{Z}$ occurring in position (q, j), there is a *special edge* from (p, i) to (q, j). Intuitively, an infinite number of values may result from the introduction of an existential variable in a given position which may lead to create another existential variable in the same position, hence an infinite number of existential variables—by imposing condition on the position dependency graph, we can avoid this from happening. This class was introduced in the context of data exchange [22].

- *Acyclic TGDs* (A): A set Σ of TGDs is *acyclic* if its position dependency graph does not contain cycles.

- *Weakly acyclic TGDs* (WA): A set Σ of TGDs is *weakly-acyclic* if its position dependency graph does not contain cycles through a special edge. Weakly acyclic TGDs enjoy several properties: they can be safely combined with EGDs, they have a finite universal model, and A $\subset$ WA.
- *Joint-acyclicity* (JA) [27]: This class arises from shifting the focus from positions to existential variables, yielding the *existential dependency graph*, where the nodes are the existential variables occurring in rules.

  Three acyclicity notions that generalize JA are *super weak acyclicity* [31] (denoted sWA), *model-faithful acyclicity* [25] (denoted MFA) and *model-summarizing acyclicity* [25] (denoted MSA), with sWA $\supset$ JA, MFA $\supset$ JA, MSA $\supset$ JA, sWA $\subset$ MFA and sWA $\subset$ MSA.
- *Acyclicity w.r.t. rule dependencies*: Another type of graph used is the *graph of rule dependencies* [8], and acyclicity in this graph gives rise to the class AGRD. Intuitively, it introduces a rule dependency relation $\succ$ (a graph of rules), for which $r_1 \succ r_2$ means that an application of rule r_1 on an instance can subsequently trigger an application of rule r_2. If the relation $\succ$ is acyclic, then no rule can trigger itself, and therefore the Skolem chase always terminates. When all strongly connected components of the graph of rule dependencies have the property of being *weakly-acyclic sets of rules* (denoted WAGRD), then the forward chaining procedure is finite.
- *Weak acyclicity after adornment and rewriting* (Adn-WA) [35] To define this class, the first step involves rewriting the set of TGDs Σ into another set Σ' by adorning the positions in the predicates that can contain infinitely many terms during the chase; if the resulting set belongs to WA, then Σ belongs to Adn-WA.
- *Hypergraph acylicity*: By requiring TGDs to have a *hypergraph-acyclic body*, the following classes arise: *body acyclic frontier guarded* [7] (ba-YG) and *body acyclic frontier one* [7] (ba-Y1).

Classes of TGDs Based on Stickiness This is an inherently different property from guardedness—informally, the main property is that variables that appear more than once in a body (join variables) are propagated (or "stick") to the inferred atoms. More formally [15], we have that *SMarking* takes a set Σ of TGDs as input and returns a set of marked TGDs where for each marked TGD, every body variable is either marked, or not. *SMarking* is defined inductively as the least fixpoint of the following:

- *Initial Marking Step.* For each TGD $\sigma \in \Sigma$ and each variable $V \in body(\sigma)$, if there exists an atom a in $head(\sigma)$ that does not contain V, mark V in σ.
- *Propagation Step.* For each TGD $\sigma \in \Sigma$ and each variable $V \in body(\sigma)$, if there exists a $\sigma' \in \Sigma$ such that $head(\sigma)$ and $body(\sigma')$ both contain a relation r and the positions where V appears in r in $head(\sigma)$ coincide exclusively with positions of marked variables in $body(\sigma')$, then mark V in σ.

We can now discuss several classes that arise from this concept.

- *Sticky Sets of TGDs* (S): A set of TGDs Σ is *sticky* if no TGD $\sigma \in SMarking(\Sigma)$ is such that a marked variable occurs in $body(\sigma)$ more than once.

- *Weakly-Sticky TGDs* (WS): Weak stickiness is a relaxation of stickiness where only "harmful" variables are taken into account. Intuitively, if a marked variable occurs more than once in a body, then at least one of these positions must be safe, since only a finite number of terms can appear in this position during the forward chaining. Note that S ⊂ WS.
- *Other sticky classes*: Sticky join TGDs, denoted SJ, generalize S to allow linear TGDs. Weakly-sticky-join TGDs, denoted WSJ, generalize both WS and SJ.

Other Classes Several other classes fall outside the two main classifications described above:

- *Full TGDs* (F): Such TGDs do not have existentially quantified variables. This property can be combined with linearity, guardedness, stickiness, and acyclicity to obtain the classes LF, GF, SF, and AF, respectively.
- *Tame TGDs* (T) [24]: *Tameness* combines the guardedness and sticky-join properties.
- *Shy TGDs* (H) [29]: A set of TGDs is *shy* if during the chase procedure nulls do not meet each other to join but only to propagate—nulls thus propagate from a single atom.

As can be seen by the number of different properties and corresponding classes reviewed here, the research community is quite active in this area. For a summary of the currently known containment relations between classes, see Fig. 1.5.

Finally, Table 1.2 summarizes the complexity results for those classes for which data, combined, *ba*-combined, and *fp*-combined complexity of conjunctive query

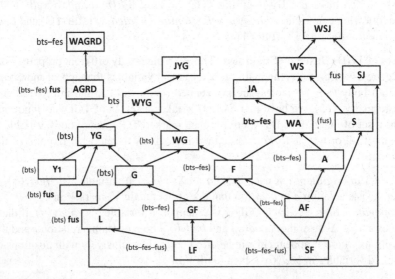

Fig. 1.5 Summary of containment relationships between different fragments of Datalog+/−; note that the fes, fus, and bts annotations also apply to all subclasses, and such inherited properties are depicted in *parentheses*

Table 1.2 Data, combined, *ba*-combined, and *fp*-combined complexity of conjunctive query answering in different languages

Fragment	Data	Combined	*ba*-combined	*fp*-combined
L	In AC_0	PSPACE-complete	NP-complete	NP-complete
LF	In AC_0	PSPACE-complete	NP-complete	NP-complete
AF	In AC_0	PSPACE-complete	NP-complete	NP-complete
G	PTIME-complete	2EXP-complete	EXP-complete	NP-complete
WG	EXP-complete	2EXP-complete	EXP-complete	EXP-complete
S	In AC_0	EXP-complete	NP-complete	NP-complete
SF	In AC_0	EXP-complete	NP-complete	NP-complete
GF	PTIME-complete	EXP-complete	NP-complete	NP-complete
F	PTIME-complete	EXP-complete	NP-complete	NP-complete
A	In AC_0	NEXP-complete	NEXP-complete	NP-complete
WA	PTIME-complete	2EXP-complete	2EXP-complete	NP-complete
WS	PTIME-complete	2EXP-complete	2EXP-complete	NP-complete
WSJ	PTIME-complete	2EXP-complete	2EXP-complete	NP-complete
T	PTIME-complete	2EXP-complete	EXP-complete	NP-complete

answering are all known. Please refer to [12, 16, 20, 22] for data complexity results and [11, 17–19, 26, 30] for combined, *ba*-combined, and *fp*-combined complexity results; furthermore, [13–15, 24] include results on all complexity types.

References

1. D. Agrawal, P. Bernstein, E. Bertino, S. Davidson, U. Dayal, M. Franklin, J. Gehrke, L. Haas, A. Halevy, J. Han, H.V. Jagadish, A. Labrinidis, S. Madden, Y. Papakonstantinou, J.M. Patel, R. Ramakrishnan, K. Ross, C. Shahabi, D. Suciu, S. Vaithyanathan, J. Widom, Challenges and opportunities with Big Data. A community white paper developed by leading researchers across the United States vol. 5 (2012), pp. 34–43
2. A. Atserias, A. Dawar, P.G. Kolaitis, On preservation under homomorphisms and unions of conjunctive queries. J. ACM **53**(2), 208–237 (2006)
3. F. Baader, D. Calvanese, D.L. McGuinness, D. Nardi, P.F. Patel-Schneider (eds.), *The Description Logic Handbook: Theory, Implementation, and Applications* (Cambridge University Press, Cambridge, 2003)
4. F. Baader, I. Horrocks, C. Lutz, U. Sattler, *An Introduction to Description Logic* (Cambridge University Press, Cambridge, 2017)
5. J. Baget, M. Mugnier, Extensions of simple conceptual graphs: the complexity of rules and constraints. J. Artif. Intell. Res. **16**, 425–465 (2002)
6. J. Baget, M. Leclaire, M. Mugnier, E. Salvat, On rules with existential variables: walking the decidability line. Artif. Intell. J. **175**(9), 1620–1654 (2011)

7. J. Baget, M. Mugnier, S. Rudolph, M. Thomazo, Walking the complexity lines for generalized guarded existential rules, in *Proceedings of the International Joint Conference on Artificial Intelligence (IJCAI)* (2011), pp. 712–717

8. J. Baget, M. Mugnier, M. Thomazo, Towards farsighted dependencies for existential rules, in *Proceedings of the Web Reasoning and Rule Systems Conference (RR)* (2011), pp. 30–45

9. C. Beeri, M.Y. Vardi, The implication problem for data dependencies, in *International Colloquium on Automata, Languages and Programming (ICALP)*, vol. 115 (1981), pp. 73–85

10. T. Berners-Lee, J. Hendler, O. Lassila, The semantic web. Sci. Am. **284**(5), 34–43 (2001)

11. A. Calì, G. Gottlob, M. Kifer, Taming the infinite chase: query answering under expressive relational constraints, in *Proceedings of the International Conference on Principles of Knowledge Representation and Reasoning (KR)* (2008), pp. 70–80

12. A. Calì, G. Gottlob, T. Lukasiewicz, A general Datalog-based framework for tractable query answering over ontologies, in *Proceedings of the ACM SIGMOD-SIGACT-SIGAI Symposium on Principles of Database Systems (PODS)* (2009), pp. 77–86

13. A. Calì, G. Gottlob, A. Pieris, Advanced processing for ontological queries. Proc. VLDB Endow. **3**(1), 554–565 (2010)

14. A. Calì, G. Gottlob, A. Pieris, Query answering under non-guarded rules in Datalog+/−, in *Proceedings of the Web Reasoning and Rule Systems Conference (RR)* (2010), pp. 1–17

15. A. Calì, G. Gottlob, A. Pieris, Towards more expressive ontology languages: the query answering problem. Artif. Intell. **193**, 87–128 (2012)

16. A. Calì, G. Gottlob, T. Lukasiewicz, A general Datalog-based framework for tractable query answering over ontologies. J.Web Semant. **14**, 57–83 (2012)

17. A. Calì, G. Gottlob, M. Kifer, Taming the infinite chase: query answering under expressive relational constraints. J. Artif. Intell. Res. **48**, 115–174 (2013)

18. A.K. Chandra, P.M. Merlin, Optimal implementation of conjunctive queries in relational data bases, in *Proceedings of the ACM Symposium on Theory of Computing (STOC)* (1977), pp. 77–90

19. A.K. Chandra, H. Lewis, J. Makowsky, Embedded implicational dependencies and their inference problem, in *Proceedings of the ACM Symposium on Theory of Computing (STOC)* (1981), pp. 342–354

20. E. Dantsin, T. Eiter, G. Gottlob, A. Voronkov, Complexity and expressive power of logic programming. ACM Comput. Surv. **33**(3), 374–425 (2001)

21. R. Fagin, P.G. Kolaitis, R.J. Miller, L. Popa, Data exchange: semantics and query answering, in *Proceedings of the International Conference on Database Theory (ICDT)* (2003), pp. 207–224

22. R. Fagin, P.G. Kolaitis, R.J. Miller, L. Popa, Data exchange: semantics and query answering. Theor. Comput. Sci. **336**(1), 89–124 (2005)

23. G. Gottlob, G. Orsi, A. Pieris, M. Šimkus, Datalog and its extensions for semantic Web databases, in *Reasoning Web International Summer School* (2012), pp. 54–77

24. G. Gottlob, M. Manna, A. Pieris, Combining decidability paradigms for existential rules. Theory Pract. Log. Program. **13**(4–5), 877–892 (2013)

25. B.C. Grau, I. Horrocks, M. Krötzsch, C. Kupke, D. Magka, B. Motik, Z. Wang, Acyclicity notions for existential rules and their application to query answering in ontologies. J. Artif. Intell. Res. **47**, 741–808 (2013)

26. D. Johnson, A. Klug, Testing containment of conjunctive queries under functional and inclusion dependencies. J. Comput. Syst. Sci. **28**(1), 167–189 (1984)

27. M. Krötzsch, S. Rudolph, Extending decidable existential rules by joining acyclicity and guardedness, in *Proceedings of the International Joint Conference on Artificial Intelligence (IJCAI)* (2011), pp. 963–968

28. M. Krötzsch, S. Rudolph, Revisiting acyclicity and guardedness criteria for decidability of existential rules. Technical Report 3011, Institute AIFB, Karlsruhe Institute of Technology (2011)

29. N. Leone, M. Manna, G. Terracina, P. Veltri, Efficiently computable Datalog programs, in *Proceedings of the International Conference on Principles of Knowledge Representation and Reasoning (KR)* (2012), pp. 13–23

30. T. Lukasiewicz, M.V. Martinez, A. Pieris, G.I. Simari, From classical to consistent query answering under existential rules, in *Proceedings of the AAAI Conference on Artificial Intelligence (AAAI)* (2015), pp. 1546–1552

31. B. Marnette, Generalized schema-mappings: from termination to tractability, in *Proceedings of the ACM SIGMOD-SIGACT-SIGART Symposium on Principles of Database Systems (PODS)* (2009), pp. 13–22

32. M. Milani, L. Bertossi, Tractable query answering and optimization for extensions of weakly-sticky Datalog+/−. arXiv:1504.03386 (2015)

33. M. Minsky, A framework for representing knowledge. Technical report, Cambridge, MA (1974)

34. M.R. Quillian, Word concepts: a theory and simulation of some basic semantic capabilities. Behav. Sci. **12**, 410–430 (1967)

35. F. Spezzano, S. Greco, Chase termination: a constraints rewriting approach. Proc. VLDB Endow. **3**(1), 93–104 (2010)

36. O. Tifrea-Marciuska, Personalised search for the social semantic web. D.Phil. Thesis, Department of Computer Science, University of Oxford (2016)

37. M.Y. Vardi, The complexity of relational query languages (extended abstract), in *Proceedings of the ACM Symposium on Theory of Computing (STOC)* (1982), pp. 137–146

Chapter 2
Models for Representing User Preferences

The study of how preferences can be modeled and leveraged in different kinds of application domains has been the subject of a wide range of works in many different disciplines. Most relevant to our work are the developments in the computer science literature, and in particular the incorporation of preferences into different kinds of query answering systems. Even within this specialized application, several approaches have been developed centered around different goals. In this chapter, we provide a brief overview of the approaches that are most relevant to our work. Before doing so, we present some basic notation that will be used here and in following chapters.

2.1 Basic Definitions and Notation

In general, we will refer to preferences over *elements*, which—as we will see below—can be either simple objects or more complex ones. In the following, we refer to a general set S of such elements, keeping their description abstract. Preferences can then be abstracted simply as binary relations representing orders; we use the symbol $\succ$ to denote such relations. Different kinds of relations arise according to what subset of the following properties they satisfy [37]:

- Reflexive: $a \succ a$, for all $a \in S$.
  If R does not satisfy this property, we say that it is *non-reflexive* (not to be confused with *irreflexive*).
- Irreflexive: $a \not\succ a$, for all $a \in S$.
- Symmetric: If $a \succ b$, then $b \succ a$, for all $a, b \in S$.
  If R does not satisfy this property, we say that it is *non-symmetric* (not to be confused with *asymmetric* or *antisymmetric*).
- Asymmetric: If $a \succ b$, then $b \not\succ a$, for all $a, b \in S$.
- Antisymmetric: If $a \succ b$ and $b \succ a$, then $a = b$, for all $a, b \in S$.

© Springer International Publishing AG 2017
G.I. Simari et al., *Ontology-Based Data Access Leveraging Subjective Reports*,
SpringerBriefs in Computer Science, DOI 10.1007/978-3-319-65229-0_2

- Transitive: If $a \succ b$ and $b \succ c$, then $a \succ c$, for all $a, b, c \in S$.
  If R does not satisfy this property, we say that it is *non-transitive* (not to be confused with *negatively transitive*).
- Negatively transitive: If $a \succ c$, then $a \succ b$ or $b \succ c$, for all $a, b, c \in S$.
- Strongly Complete: $a \succ b$ or $b \succ a$, for all $a, b \in S$.
- Complete: $a \succ b$ or $b \succ a$, for all $a, b \in S$ such that $a \neq b$.

Clearly, these properties are not completely independent of each other; for instance, if both transitivity and symmetry hold, then so does reflexivity (if $a \succ b$ and $b \succ a$, then $a \succ a$). There are therefore different characterizations for the different kinds of relations that usually appear in the literature. The main ones are the following [37]:

- **Order:** transitive and antisymmetric.
- **Partial Order:** reflexive, transitive, and antisymmetric.
- **Strict Partial Order:** transitive and asymmetric. Intuitively, the relation can be represented by a directed acyclic graph.
- **Linear Order:** transitive, asymmetric, and strongly complete. Intuitively, all elements in the set can be sorted in ascending or descending order. These relations are also sometimes referred to as *total* orders.
- **Strict Linear Order:** transitive, asymmetric, and complete.
- **Weak Order:** transitive and strongly complete. Intuitively, these orders are like linear orders that allow "ties", and all pairs of elements that are "tied at the same level" are in the relation.
- **Strict Weak Order:** asymmetric and negatively transitive. Intuitively, these orders are like linear orders that allow "ties".

To better understand how this naming convention works, we can informally say that *strict* refers to irreflexive relations, *weak* refers to allowing "ties", and *partial* means that there are incomparable pairs of elements. A graphical representation of the different kinds of orders is shown in Fig. 2.1.

In the rest of this chapter (and book in general), we will assume that a "*preference relation*" (denoted with the symbol $\succ$, as above), is defined over a set S of elements and is a strict partial order (SPO) over S (so, a relation that can be visualized as a directed acyclic graph)—these are generally considered to be the minimal requirements for a reasonable preference relation. If $a \succ b$, we say that a is *preferred to b*. The *indifference relation* induced by $\succ$, denoted with $\sim$, is defined as follows: for any $a, b \in S, a \sim b$ iff $a \nsucc b$ and $b \nsucc a$.

A *stratification* of S relative to $\succ$ is an ordered sequence $S_1, \ldots, S_k$, where each S_i is a maximal subset of S such that for every $a \in S_i$, there is no $b \in \bigcup_{j=i}^{k} S_j$ with $b \succ a$. Intuitively, S_1 contains the most preferred elements in S relative to $\succ$; then, S_2 contains the most preferred elements of $S - S_1$, and so on. Stratifications always exist, are unique, and are a partition of S. Elements in stratum S_i have *rank i*. The rank of an element $a \in S$ relative to $\succ$ is denoted as $rank(a, \succ)$.

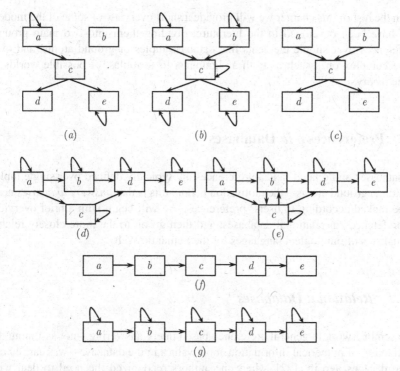

Fig. 2.1 A graphical depiction of the different kinds of orders, where *transitive edges are implicit*. (**a**) Order; (**b**) Partial order; (**c**) Strict partial order; (**d**) Weak order; (**e**) Strict weak order; (**f**) Strict linear order; and (**g**) Linear order

Qualitative vs. Quantitative Preferences Models One of the most basic distinctions between preference models and the underlying preference relations that they represent is that of *qualitative* vs. *quantitative*. Essentially, quantitative models are equivalent to the assignment of a numerical score to each element in the set of interest—depending on whether ties are allowed or not, this gives rise to either a strict weak order or a linear order, respectively.

On the other hand, qualitative models allow preferences to be expressed by referring to different aspects of the elements, such as location and size in the case of apartments. Since there may exist, for instance, a large apartment in a bad location and another one that is small but in a good location, the result can be a relation with incomparable elements. Clearly, qualitative models are richer than quantitative ones, since the relations that the former are capable of representing strictly contain those of the latter.

Individual vs. Group Preferences Another important aspect that distinguishes approaches to preference modeling is whether they aim to represent single users or groups. See Sect. 2.4 for a brief discussion and references to relevant literature on this aspect.

In the rest of this chapter we will provide a short overview of some of the models that have been developed in the literature, dividing them into two main groups: preferences over simple elements like database tuples or ground atoms, and over more complex ones such as truth assignments to formulas, or possible worlds in some theory.

2.2 Preferences *à la* Databases

In databases and related fields, preferences are typically defined over single tuples or atomic ground atoms—the motivation behind this is that *answers to queries* need to be ranked according to users' preferences. We will begin with a brief overview of preferences in relational databases, and then go on to the more closely related formalisms of ontological languages for the Semantic Web.

2.2.1 Relational Databases

The seminal work in general preference-based query answering—not assuming the availability of numerical information about values in the database—was carried out three decades ago in [22], where the authors recognized the need to deal with situations in which a given query either has no answers or too many—in the first case, conditions can be weakened, while in the latter they can be strengthened. Their solution was to extend the SQL language to incorporate user preferences via a new PREFER clause that can be either simple or compound, as well as nested for richer combinations. The expressivity of the resulting language is determined via a mapping to the domain relational calculus [23], and the authors briefly comment on a prototype Prolog implementation. This work subsequently inspired an entire line of research, as well as implemented systems. For instance [21], describes the Preference SQL system, which has been commercially available since 1999.

The Skyline Operator An important development in this line of research is the well-known *skyline* operator, which was first introduced in [5] as a way to characterize *interesting points*—essentially, a point (a tuple, in this setting) is considered to be interesting, if it is not dominated by any other point. When preferences are specified in terms of multiple aspects, such as location and price for a restaurant, then a dominance relation can be simply defined to hold between two points whenever the dominant one has a better value (or at least one that is just as good) than the dominated one in *every* aspect. The term "skyline" arises from the visualization of a cityscape—the points that form the skyline in a given dataset resemble the buildings that stand out against the sky because they are either very tall or they are closer than the others to the viewpoint (in this example, the dimensions are thus height and distance to the viewpoint). Figure 2.2 shows a simple illustration

Fig. 2.2 Set of restaurants plotted with respect to their evaluation in terms of two dimensions of interest to a hypothetical user: *Location* and *Food*; both metrics are assumed to be defined so that higher values are better. The *circled points* comprise the skyline

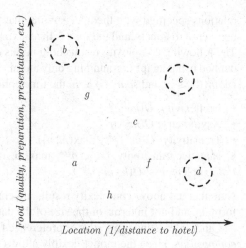

of this concept: restaurants are evaluated in terms of location with respect to the user's hotel and food (for instance, according to the restaurant's average "Food" rating on TripAdvsior). Restaurants *b*, *e*, and *d* comprise the skyline; *b* has a great rating in terms of food, though it is quite bad in terms of location; *d* is somewhat the opposite, since it is among the worst in terms of food, but very close to the hotel; finally, *e* clearly dominates all non-skyline points.

In [5], the authors develop an extension of SQL with an optional SKYLINE OF clause in which users can express their desire to either minimize or maximize different dimensions (table attributes), or that values should be different in the result. Seven algorithms are provided for the physical implementation of the new clause (based on block nested loops and divide and conquer approaches, plus a special one that only works for two-dimensional cases). Skyline queries can also be implemented directly in standard SQL, though performance suffers greatly.

Skylines play an important role in preference-based query answering systems; in general, they are used in combination with top-*k* answers to yield the user's preferred answers.

Preference Formulas In [11], the *preference formula* formalism was introduced as a way to flexibly specify qualitative preferences, as well as embed them into queries. Preference formulas have the form:

$$t_1 \succ_C t_2 \ \text{iff} \ C(t_1, t_2),$$

where t_1 and t_2 are database tuples, and $C(t_1, t_2)$ is a first-order formula that may contain equality (or inequality) constraints and/or rational-order constraints (equality, inequality, and less/greater than comparisons with rational numbers). A more flexible variant was also used in [26], where "if" is used instead of "iff" in the definition. Chomicki refers to formulas with only built-in predicates as *intrinsic preference formulas* (ipfs), and studies the complexity of checking properties of relations specified with them. Assuming that formulas are in DNF, the *width* is the number of disjuncts, and the *span* is the maximum number of conjuncts in a disjunct. The following complexity results hold in this case [11]; given a preference relation defined using an ipf C containing only atomic constraints over a single domain, with $width(C) \leq m$ and $span(C) \leq n$, the time complexity of checking properties is:

- Irreflexivity: $O(m \cdot n)$
- Asymmetry: $O(m^2 \cdot n)$
- Transitivity: $O(m^2 \cdot n^m \cdot \max(m, n))$.
- Negative transitivity: $O(m \cdot n^{2m} \cdot \max(m, n))$
- Completeness: $O(k \cdot m \cdot n^{2m})$

Note that the above complexity results are characterized in terms of the size of the formula, and not in terms of the size of the database.

Other interesting properties of preference formulas are those resulting from their *composition*, since the model flexibly allows their combination via operators, such as union, intersection, and difference. Interestingly, neither weak nor total orders are preserved by such operations; on the other hand, strict partial orders are preserved by intersection, though not by union or difference.

More complex composition operations can also be defined [11]:

- Prioritized composition: Prefer according to $\succ_2$, unless $\succ_1$ is applicable:

$$t_1 \succ_{1,2} t_2 \equiv t_1 \succ_1 t_2 \vee (t_1 \sim_1 t_2 \wedge t_1 \succ_2 t_2).$$

This kind of composition is associative and distributes over union. Weak and total orders are preserved, but strict partial orders are not.

- Pareto composition: Defined over the Cartesian product of two relations:

$$(t_1, t_2) \succ_P (t_1', t_2') \equiv (t_1 \succeq_1 t_1') \wedge (t_2 \succeq_2 t_2') \wedge \left((t_1 \succ_1 t_1') \vee (t_2 \succ_2 t_2') \right).$$

Pareto composition does not preserve strict partial order, weak order, or total order.

- Lexicographic composition: Defined over the Cartesian product of two relations:

$$(t_1, t_2) \succ_L (t_1', t_2') \equiv (t_1 \succeq_1 t_1') \vee (t_1 \sim_1 t_1') \wedge (t_2 \succ_2 t_2').$$

This kind of composition preserves weak and total orders, but not strict partial ones.

The *winnow* operator is defined as a companion to a preference formula C over a database instance r as follows:

$$\omega_C(r) = \{t \in r \mid \nexists t' \in r \text{ such that } t' \succ_C t\}.$$

Clearly, the winnow operator characterizes the skyline of a database instance with respect to a preference relation, as discussed above; the main difference is that skylines as proposed in [5] are defined for single relations. Several algebraic properties of this operator are investigated in [11], such as commutativity, commutativity with selection, commutativity with projection, distribution over Cartesian product, and distribution over union and difference.

Other Notes Studies of preferences related to (active) databases have also been done in classical logic programming [18, 19] as well as answer set programming frameworks [7]. For a fairly recent survey of preference-based query answering formalisms in databases, we refer the reader to [41].

2.2.2 Ontology Languages

There have been only a few approaches to adding preferences to ontology languages. The first one—to our knowledge—was [39], where an extension is developed so that users can add their preferences to SPARQL queries via a new PREFERRING solution sequence modifier that allows to return either just skyline answers or soft constraints (where preference is given to answers that satisfy them, but they can also be relaxed if necessary). Other early approaches for preference-based querying RDF graphs include [8, 9, 32].

The approach that is most closely related to the one used in this book is the PrefDatalog+/− formalism, which was first presented in [26]; later, it was extended to work in combination with preference models based on probability values [30], as well as groups of users [29]. We will now provide a brief overview, starting with a toy ontology that will be used as a running example.

Example 2.1 Consider the following simple ontology $O = (D, \Sigma)$ describing gift ideas for friends:

$$D = \{\text{scifi-book}(b_1, \text{asimov}), \text{scifi}(b_2, \text{asimov}), \text{fantasy-book}(b_3, \text{tolkien}),$$
$$\text{puzzle}(p_1), \text{mmorpg}(v_1), \text{actionAdventure}(v_2)\}.$$

$$\Sigma = \{\text{scifi-book}(T, A) \rightarrow \text{book}(T, A),$$
$$\text{fantasy-book}(T, A) \rightarrow \text{book}(T, A),$$
$$\text{mmorpg}(X) \rightarrow \text{videoGame}(X),$$
$$\text{actionAdventure}(X) \rightarrow \text{videoGame}(X),$$
$$\text{book}(T, A) \rightarrow \text{educational}(T),$$
$$\text{puzzle}(N) \rightarrow \text{educational}(N),$$
$$\text{book}(X) \wedge \text{videoGame}(X) \rightarrow \bot\}.$$

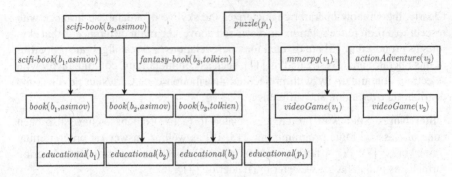

Fig. 2.3 Chase graph for the ontology from Example 2.1—*arrows* indicate TGD applications; nodes without incoming edges are part of the database

This ontology specifies a taxonomy with two kinds of book genres (science fiction and fantasy) and two video game genres (massively multi-player online role playing and action/adventure); it also states that books and puzzles are categorized under an "educational" label. The database D contains several instances for each of these genres. Finally, the last formula is a negative constraint stating that an item cannot be both a book and a video game.

Figure 2.3 shows the chase graph that arises from applying the TGDs in Σ over the database D.   ∎

Though the PrefDatalog+/− formalism does not commit to a specific preference framework, in this overview, we adopt the *preference formulas* approach discussed above. Consider the following example.

Example 2.2 Continuing with Example 2.1, consider the following formulas representing a specific user's preferences for children's gifts:

C_1: *educational*$(X) \succ$ *videoGame*(Y) if $\top$;
C_2: *book*$(T_1, A_1) \succ$ *book*(T_2, A_2) if *scifi_book*$(T_1, A_1) \wedge$ *fantasy_book*(T_2, A_2);
C_3: *book*$(T_1, A_1) \succ$ *book*(T_2, A_2) if $(T_1 = b_1) \wedge (T_2 = b_2)$;
C_4: *educational*$(X) \succ$ *educational*(Y) if *puzzle*$(X) \wedge$ *fantasy_book*(Y);
C_5: *videoGame*$(X) \succ$ *videoGame*(Y) if *actionAdventure*$(X) \wedge$ *mmorpg*(Y).

The formula C_1 states that educational gifts are preferable to video games (unconditionally), C_2 states that sci-fi books are preferred over fantasy ones, C_3 refers specifically to books b_1 and b_2 (the former is preferred), C_4 states that an educational gift that is a puzzle is preferable to one that is a fantasy book, and C_5 specifies preference of action/adventure video games over massively multiplayer online role playing ones.   ∎

In the following, for a preference formula $C : t_1 \succ t_2$ if $C(t_1, t_2)$, we call $C(t_1, t_2)$ the *condition* of C, and denote it with $cond(C)$.

Syntax and Semantics of PrefDatalog+/− As discussed in Chap. 1, for classical Datalog+/− we have an infinite universe of constants Δ_{Ont}, an infinite set of variables $\mathcal{V}_{Ont}$, and a finite set of predicate names $\mathcal{R}_{Ont}$. Analogously, for the preference model, we have a finite set of constants Δ_{Pref}, an infinite set of variables $\mathcal{V}_{Pref}$, and a finite set of predicate names $\mathcal{R}_{Pref}$. In the following, we assume w.l.o.g. that $\mathcal{R}_{Pref} \subseteq \mathcal{R}_{Ont}$, $\Delta_{Pref} \subseteq \Delta_{Ont}$, and $\mathcal{V}_{Pref} \subseteq \mathcal{V}_{Ont}$. These sets give rise to corresponding *Herbrand bases* consisting of all possible ground atoms that can be formed, which we denote by $\mathcal{H}_{Ont}$ and $\mathcal{H}_{Pref}$, respectively. Clearly, we have $\mathcal{H}_{Pref} \subseteq \mathcal{H}_{Ont}$, meaning that preference relations are defined over a subset of the possible ground atoms.

Let O be a Datalog+/− ontology and P be a set of preference formulas with Herbrand bases $\mathcal{H}_{Ont}$ and $\mathcal{H}_{Pref}$, respectively. A *preference-based Datalog+/−ontology* (PrefDatalog+/− ontology, or knowledge base) is of the form $KB = (O, P)$, where $\mathcal{H}_{Pref} \subseteq \mathcal{H}_{Ont}$. The semantics of PrefDatalog+/− arises as a direct combination of the semantics of Datalog+/− and that of preference formulas. A knowledge base $KB = (O, P)$ *satisfies* $a_1 \succ_P a_2$, denoted $KB \models a_1 \succ_P a_2$, if and only if:

1. $O \models a_1$ and $O \models a_2$, and
2. $\models \bigvee_{pf_i \in P} cond(pf_i)(a_1, a_2)$.

Intuitively, the consequences of KB are computed in terms of the chase for the classical Datalog+/− ontology O, and the set of preference formulas P describes the preference relation over pairs of atoms in $\mathcal{H}_{Ont}$. Considering $KB = (O, P)$, where O is the ontology, and P is the set of preference formulas from the running example, we have that:

$$KB \models educational(b_1) \succ_P videoGame(v_1),$$

since $O \models educational(b_1)$, $O \models videoGame(v_1)$, and $C_1 \in P$ allows us to conclude that $educational(b_1) \succ_P videoGame(v_1)$. On the other hand,

$$KB \not\models educational(b_1) \succ_P educational(b_2),$$

since the second condition is not satisfied in this case.

Preference-Based Queries There are two kinds of preference-based queries that can be issued over PrefDatalog+/− ontologies: *skyline* and *k-rank*. Answers to such queries are defined as usual via *substitution* (functions from variables to variables or constants) and *most general unifiers* [25]. We consider two kinds of classical queries: disjunctive atomic queries (disjunctions of atoms—DAQs) and conjunctive queries (CQs). We begin with the former: let $Q(\mathbf{X}) = q_1(\mathbf{X}_1) \vee \cdots \vee q_n(\mathbf{X}_n)$, where the q_i's are atoms and $\mathbf{X}_1 \cup \cdots \cup \mathbf{X}_n = \mathbf{X}$:

- The set of *skyline answers* to Q is defined as:

$$\{\theta q_i \mid O \models \theta q_i \text{ and } \nexists \theta' \text{ such that } O \models \theta' q_j \text{ and } \theta' q_j \succ_P \theta q_i, \text{ with } 1 \leqslant i, j \leqslant n\},$$

where θ, θ' are most general unifiers for the variables in $Q(\mathbf{X})$.

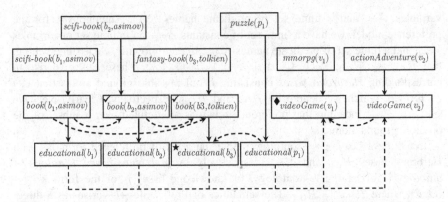

Fig. 2.4 Chase graph from Fig. 2.3, augmented with information on preferences between atoms based on the preference formulas from Example 2.2—*dashed arrows* denote the preference relation that arises from that model. *Marks on the upper left-hand corner* indicate dominance marks associated with different queries

- A *k-rank answer* to $Q(\mathbf{X})$ is defined for transitive relations $\succ_P$ as a sequence of maximal length of mgu's for $\mathbf{X}$: $S = (\theta_1, \ldots, \theta_{k'})$ such that $O \models \theta_i Q$ for $1 \leq i \leq k' \leq k$, and S is built by subsequently appending the skyline answers to Q, removing these atoms from consideration, and repeating the process until either $S = k$ or no more answers to Q remain.

Note that k-rank answers are only defined when the preference relation is transitive; this kind of answer can be seen as a generalization of traditional top-k answers [41] that are still defined when $\succ_P$ is not a weak order, and their name arises from the concept of *rank* introduced in [11].

Intuitively, for DAQs, both kinds of answers can be seen as atomic consequences of O that satisfy the query: the skyline answers can be seen as sets of atoms that are not dominated by any other such atom, while k-rank answers are k-tuples sorted according to the preference relation. We refer to these as *answers in atom form*.

Returning to the running example, consider the queries:

$$Q_1(X, Y) = book(X, Y) \quad \text{and} \quad Q_2(X) = educational(X).$$

The $\succ_P$ relation is depicted in Fig. 2.4 (the arcs with dashed lines denote the ordered pairs in the relation). The set of skyline answers to Q_1 is: $\{book(b_1, asimov)\}$, while for Q_2, it is $\{educational(p_1), educational(b_1), educational(b_2)\}$. A 3-rank answer to Q_1 is:

$$(book(b_1, asimov), book(b_2, asimov), book(b_3, tolkien)).$$

For Q_2, a 3-rank answer is

$$(educational(p_1), educational(b_1), educational(b_2)).$$

Finally, the query

$$Q_3 = puzzle(X) \vee videoGame(X)$$

yields $\{puzzle(p_1), videoGame(v_2)\}$ as skyline answers, and a 3-rank answer is

$$\{puzzle(p_1), videoGame(v_2), videoGame(v_1)\}.$$

In the case of (non-atomic) conjunctive queries, the substitutions in answers no longer yield single atoms but rather *sets* of atoms. Therefore, to answer such queries relative to a preference relation, we must extend the preference specification framework to take into account sets of atoms instead of individual ones. One such approach was proposed in [45], where a mechanism to define a preference relation over tuple sets $\succ_{PS}: 2^{\mathcal{H}_{Pref}} \times 2^{\mathcal{H}_{Pref}}$ is introduced. We will not discuss this further here; [26] includes a treatment of their complexity and briefly describes how methods from the relational databases literature can be applied to answer them.

The Preference-Augmented Chase (*prefChase*) To compute skyline and k-rank answers to queries over a PrefDatalog+/− ontology $KB = ((D, \Sigma), P)$, an *augmented chase forest* is used, which is comprised of the necessary finite part of the chase forest relative to a given query that is augmented with an additional kind of edge called *preference edges*—these occur between nodes labeled with $a, b \in chase(D, \Sigma)$ if and only if $a \succ_P b$. Finally, when an edge is introduced *between nodes whose labels satisfy* Q, the node with the incoming edge is *marked*. Figure 2.4 shows *prefChase(KB, Q)* for the PrefDatalog+/− ontology from the running example; for illustrative reasons, markings for three different queries have been included in the figure: $book(X, Y)$ (check mark), $educational(X)$ (star), and $videoGame(X)$ (diamond).

This data structure can directly be used to answer queries. Obtaining the node markings consists of almost all of the work towards answering a skyline query; all that remains to be done is to go through the structure and find the nodes whose labels satisfy the query and, if unmarked, add them to the output. For k-rank queries, the query answering process involves iterating through the computation of the skyline answers, updating the result by appending these answers in arbitrary order, and removing the nodes and edges involved in the result from the chase structure; finally, before the next iteration, the node markings need to be updated.

As an upper bound on the cost of these procedures, both kinds of queries can be answered in time quadratic in the cost of building the classical chase (in the data complexity) [26]—this cost depends on the fragment of Datalog+/− used in the underlying ontology, but the overhead imposed by the markings and preference edges to obtain the preference-augmented chase is clearly at most quadratic.

2.2.3   Models for Uncertain Preferences

There are several ways in which uncertainty can be incorporated into models for preference-based reasoning in the databases tradition. We will provide a brief overview of the two that are most relevant to this work.

PP-Datalog+/− The first model we will review is called *Probabilistic Preference-based Datalog+/−* [27, 30]. It is an extension of PrefDatalog+/−, where the main idea is to combine it with the probabilistic extension proposed in [17]; for the purposes of this discussion, the only important aspect of the latter is that any atomic inference can be assigned a probability with which it holds—such probability values naturally induce a weak order over the elements, under the assumption that more probable inferences are more preferable in that model.

The presence of two preference models—an SPO provided by the user and a weak order induced by the probability values—then poses the problem of how to order the results of a query. Figure 2.5 illustrates this situation in the domain of Example 2.1. As usual, the SPO represents the user's preferences, while the probability values could indicate, for instance, the probability that each book will be delivered in time. Note that books b_2 and b_4, which are the top two picks, have the

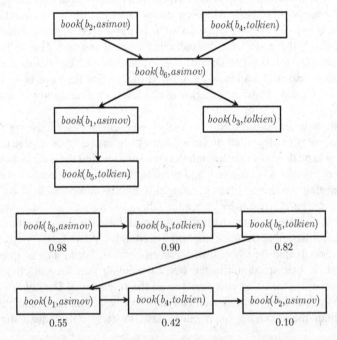

Fig. 2.5 *Top*: Strict partial order provided by the user; *Bottom*: Weak order arising from the probability values shown (under the assumption that more probable inferences are preferred in that model). For clarity, edges arising from the transitive closure are not shown

lowest probabilities; in this case, it seems that the user would be best off choosing book b_6, which is the next preferred and has a very high probability of reaching them in time.

In order to formalize this, the concept of *preference combination operators* is introduced in [30]. The goal of such operators is to produce a new preference relation that satisfies a set of basic properties. Two classes of operators are proposed: (1) *Egalitarian* operators allow the resulting relation to eventually resemble either the SPO or the probability-based orders; (2) *User-biased* operators, on the other hand, base the resulting preference relation on the user's preferences, and use the probabilistic model as a secondary source of "advice". Of the specific operators defined for each family in [30], though the theoretical complexity of the algorithms is similar, the actual properties of specific inputs—such as edge density, size of skylines in the SPO, and number of ties in the weak order—greatly affect their running time in practice.

PPLNs The other model that we would like to mention here is that of *Probabilistic Preference Logic Networks* [28], which is also known as *Markov Models over Weighted Orderings* [31]. Like PP-PrefDatalog+/−, PPLNs combine user's preferences with probabilistic uncertainty; however, this is done in a very different way: users provide preference formulas that are Boolean combinations of atomic statements of the form "$a \succ b$", and each formula receives a *weight*. This allows us to model situations that are characterized by: (1) the fact that we only have information on certain pairs of elements; and (2) the uncertainty underlying the information provided—users are much more likely to express preferences that are subject to exceptions than ones that hold all the time.

The semantics of PPLNs is based on *Markov Random Fields* (MRFs), a classical model for representing distributions over possible worlds; instead of having possible worlds arising from truth assignments to Boolean variables, possible worlds in the PPLN approach are *linear orderings*, so their number is factorial in the number of elements instead of exponential. Computing the probability of a preference statement issued as a query to a PPLN has been shown to be #P-hard [28], so two approaches based on results from the mathematical branch of order theory are investigated to tackle intractability: approximations via approximate model counting (yielding a fully polynomial randomized approximation scheme under fixed-parameter assumptions) and exact approaches via exact counting that work for specific fragments (yielding a fixed-parameter polynomial time algorithm).

Another approach to tractable query answering in these models was proposed in [31], where variable elimination algorithms inspired by those for related models like Bayesian networks were investigated and empirically evaluated. For *chain* models (in which weighted preference statements taken as a whole form sets of concatenated formulas), a cubic time algorithm is proposed and empirically shown to scale well—queries can be answered in under a minute for models of up to 2500 elements. For the more general case of models consisting of weighted atomic statements (and queries/evidence consisting of conjunctions of such formulas), an algorithm is proposed that, although of exponential running time, has the *linear cut size* of the model (a measure similar to treewidth) as dominating factor. Empirical evaluations of this algorithm show that tractability is much more elusive than for chain models; however, even simple heuristics based on minimizing linear cut size have a large impact both on running time and the size of the data structures that must be maintained.

2.3   Preferences *à la* Philosophy and Related Disciplines

Preferences in philosophy have received much attention, as far back as Aristotle, perhaps because they constitute a purely subjective evaluation of the elements in question. In contrast to the approach taken in databases as described in the previous section, the tradition in philosophy and related disciplines is to consider preferences over elements that are mutually exclusive—the most common sets of elements, especially in disciplines borne out of philosophy, such as logic and decision theory, are *states of the world*, which can be represented as vectors of values for a set of variables. For instance, one might prefer sunny weather on weekends over sunny weather on weekdays. Note that, in general, this leads to large sets of alternatives, since these are comprised of the Cartesian product of the domains of all variables. This is quite different from the simpler view taken in databases and related approaches, where atomic elements are typically (but not always) referred to in preference relations. In contrast, the philosophical tradition does not exclude reasoning about atomic elements—the separation implied by dividing the discussion into two separate sections of this chapter is merely a practical one.

Since this conception of preference modeling is somewhat less related to the tools developed in this book, here, we will only review the *CP-nets* model.

CP-Nets This model, first proposed in [6], is designed to represent preferences over possible worlds much in the same way as probabilities are represented in Bayesian networks. Their name comes from either *conditional preference* or *ceteris paribus* (meaning "all else being equal")—variables have corresponding nodes in a directed graph, and conditional tables state preferences between the different values that they can take, given values for the variables that are directly connected by incoming edges in the graph. The set of *outcomes* is comprised of all possible settings of the variables in the model. The semantics of CP-nets is given by the concept of *worsening flip* between outcomes; given two outcomes that differ in only one value, the values that remain fixed allow the ordering of the pair according to the relevant conditional preference table(s)—this is where "ceteris paribus" comes into play. An outcome is said to *dominate* another, if there is a sequence of worsening flips from the first to the second.

The two main reasoning tasks associated with CP-nets are: deciding whether one outcome dominates another (*dominance query*), and deciding whether a given outcome is optimal (*outcome optimization*). The following example illustrates these concepts.

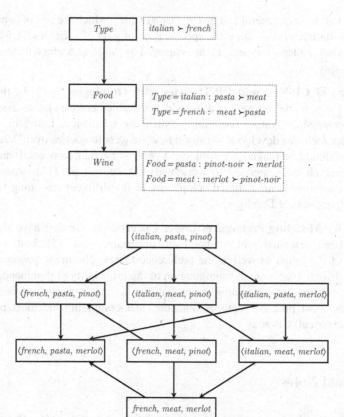

Fig. 2.6 *Top*: A simple CP-net specifying preferences over restaurant types, foods, and wine. *Bottom*: Graph of all possible outcomes; edges denote preference between outcomes

Example 2.3 Consider the toy example CP-net depicted in Fig. 2.6 (top), containing three variables relevant to the restaurant domain: restaurant type, foods, and wine—all variables are binary in order to keep the example simple.

According to this CP-net, the user prefers Italian restaurants over French ones; if they are in an Italian restaurant, then pasta is preferred over meat, and the other way around, if the restaurant is French. Finally, when having pasta, Pinot Noir is preferred over Merlot, whereas the opposite holds when having meat.

Figure 2.6 (bottom) specifies all possible worsening flips between outcomes, from which the full preference relation can be obtained. For example, outcome ⟨*italian, meat, merlot*⟩ is preferred over ⟨*italian, meat, pinot*⟩, because of the first entry in the conditional preference table for the *Food* variable, which specifies that when having meat, Merlot is preferred over Pinot Noir.

For this model, note that outcome ⟨*italian, pasta, merlot*⟩ dominates outcome ⟨*french, meat, merlot*⟩ since there is a sequence of worsening flips of length two that goes through ⟨*italian, meat, merlot*⟩. Finally, outcome ⟨*italian, pasta, pinot*⟩ is optimal, since there is no other outcome that dominates it. ∎

CP-nets were later generalized by *CP-theories* [43], which are sets of conditional preference statements that allow the specification of a set of variables for which the value does not matter—CP-nets are thus captured by the case in which these variable sets are empty.

Extensions of CP-Nets and CP-Theories with Ontologies In [13], the *onto-logical CP-net* model is presented, in which variables correspond to description logic axioms whose values are simply satisfied/not satisfied. Later, in [14], a similar approach was developed to inform how answers to queries over Datalog+/− ontologies should be ranked, focusing on skyline and *k*-rank answers. Finally, this line of research was extended to work with CP-theories in [15], studying the data and combined computational complexity of the different reasoning tasks for different fragments of Datalog+/−.

CP-Nets for Modeling Preferences Under Uncertainty CP-nets have also more recently been extended with probabilistic uncertainty [4, 12], both over the structure of the graph as well as the preference tables. The main reasoning tasks in probabilistic CP-nets are the computation of the probability of dominance of one outcome over another, computing the probability that a given outcome is optimal, finding the most probable optimal outcome, and computing the most probable induced (classical) CP-net.

2.4   Final Notes

Preferences and Groups The area of modeling different kinds of groups is quite related to the study of preferences. For instance, *social choice* focuses on mechanisms for finding the decision that is best for a group as a whole by combining the opinion of individuals; this has been the topic of study in different fields, like mathematics, economics, politics, and sociology for decades [35, 42]. Other areas related to social choice are multiagent systems [44], recommender systems [2, 34, 40], rank aggregation [16], and combining incomplete preferences [1, 24, 36]. Finally, there is also an approach based on Datalog+/− for query answering with group preferences that was recently proposed in [29].

Preferences and Provenance This work is also closely connected to the study and use of *provenance* in information systems and, in particular, the Semantic Web and social media [3, 33]—provenance refers to the description of the history of data in its life cycle, and it is also sometimes referred to as *lineage*. Research in provenance distinguishes between *data* and *workflow* provenance: the former explores the data flow within applications in a fine-grained way, while the latter is coarse-grained and does not consider the flow of data. Another classification is typically considered in the provenance literature is the *why*, *how*, and *where* framework [10]. As we will see in Chap. 3, our data models incorporate provenance information via registers that store information about the origin of each report, allowing us to take into account where evaluations within a social media system come from (such as information about who has issued the report, their origin, and what their preferences were at the time), and leverage this information to allow users to make informed provenance-based decisions.

For a more comprehensive review of the many aspects associated with reasoning with preferences, the interested reader is referred to [38] and [20].

References

1. M. Ackerman, S. Choi, P. Coughlin, E. Gottlieb, J. Wood, Elections with partially ordered preferences. Public Choice **157**(1/2), 145–168 (2013)
2. S. Amer-Yahia, S.B. Roy, A. Chawlat, G. Das, C. Yu, Group recommendation: semantics and efficiency. Proc. VLDB Endow. **2**(1), 754–765 (2009)
3. G. Barbier, Z. Feng, P. Gundecha, H. Liu, *Provenance Data in Social Media* (Morgan and Claypool, San Rafael, CA, 2013)
4. D. Bigot, B. Zanuttini, H. Fargier, J. Mengin, Probabilistic conditional preference networks, in *Proceedings of the Conference on Uncertainty in Artificial Intelligence UAI* (2013)
5. S. Börzsönyi, D. Kossmann, K. Stocker, The Skyline operator, in *Proceedings of the International Conference on Data Engineering (ICDE)* (2001), pp. 421–430
6. C. Boutilier, R.I. Brafman, C. Domshlak, H.H. Hoos, D. Poole, CP-nets: a tool for representing and reasoning with conditional ceteris paribus preference statements. J. Artif. Intell. Res. **21**, 135–191 (2004)
7. G. Brewka, Preferences, contexts and answer sets, in *Proceedings of the International Conference on Logic Programming (ICLP)* (2007), p. 22
8. G. Chamiel, M. Pagnucco, Exploiting ontological information for reasoning with preferences, in *Multidisciplinary Workshop on Advances in Preference Handling* (2008)
9. L. Chen, S. Gao, K. Anyanwu, Efficiently evaluating skyline queries on RDF databases, in *The Semantic Web: Research and Applications* (2011), pp. 123–138
10. J. Cheney, L. Chiticariu, W. Tan, Provenance in databases: why, how, and where. Found. Trends Databases **1**(4), 379–474 (2009)
11. J. Chomicki, Preference formulas in relational queries. ACM Trans. Database Syst. **28**(4), 427–466 (2003)
12. C. Cornelio, J. Goldsmith, N. Mattei, F. Rossi, K.B. Venable, Dynamic probabilistic CP-nets, in *Proceedings of the Multidisciplinary Workshop on Advances in Preference Handling* (2013), pp. 1–7
13. T. Di Noia, T. Lukasiewicz, G.I. Simari, Reasoning with semantic-enabled qualitative preferences, in *Proceedings of the International Conference on Scalable Uncertainty Management (SUM)* (2013), pp. 374–386
14. T. Di Noia, T. Lukasiewicz, M.V. Martinez, G.I. Simari, O. Tifrea-Marciuska, Computing k-rank answers with ontological CP-nets, in *Proceedings of the Workshop on Logics for Reasoning about Preferences, Uncertainty, and Vagueness (PRUV)* (2014), pp. 74–87
15. T. Di Noia, T. Lukasiewicz, M.V. Martinez, G.I. Simari, O. Tifrea-Marciuska, Combining existential rules with the power of CP-Theories, in *Proceedings of the International Joint Conference on Artificial Intelligence (IJCAI)* (2015), pp. 2918–2925
16. R. Fagin, R. Kumar, D. Sivakumar, Comparing top k lists. SIAM J. Discret. Math. **17**(1), 134–160 (2003)
17. G. Gottlob, T. Lukasiewicz, M.V. Martinez, G.I. Simari, Query answering under probabilistic uncertainty in Datalog+/– ontologies. Ann. Math. Artif. Intell. **69**, 37–72 (2013)
18. K. Govindarajan, B. Jayaraman, S. Mantha, Preference logic programming, in *Proceedings of the International Conference on Logic Programming (ICLP)* (1995), pp. 731–745

19. K. Govindarajan, B. Jayaraman, S. Mantha, Preference queries in deductive databases. N. Gener. Comput. **19**(1), 57–86 (2001)
20. S. Kaci, in *Working with Preferences: Less Is More*. Cognitive Technologies (Springer, Berlin/Heidelberg, 2011)
21. W. Kießling, M. Endres, F. Wenzel, The preference SQL system: an overview. IEEE Data Eng. Bull. **34**(2), 11–18 (2011)
22. M. Lacroix, P. Lavency, Preferences: putting more knowledge into queries, in *Proceedings of the International Conference on Very Large Databases (VLDB)* (1987), pp. 217–225
23. M. Lacroix, A. Pirotte, ILL: an English structured query language for relational data bases. ACM SIGART Bull. (61), 61–63 (1977), http://dl.acm.org/citation.cfm?id=1045335
24. J. Lang, M.S. Pini, F. Rossi, D. Salvagnin, K.B. Venable, T. Walsh, Winner determination in voting trees with incomplete preferences and weighted votes. J. Auton. Agent. Multi-Agent Syst. **25**(1), 130–157 (2012)
25. J.W. Lloyd, *Foundations of Logic Programming*, 2nd edn. (Springer, Berlin/Heidelberg, 1987)
26. T. Lukasiewicz, M.V. Martinez, G.I. Simari, Preference-based query answering in Datalog+/– ontologies, in *Proceedings of the International Joint Conference on Artificial Intelligence (IJCAI)* (2013), pp. 1017–1023
27. T. Lukasiewicz, M.V. Martinez, G.I. Simari, Preference-based query answering in probabilistic Datalog+/– ontologies, in *Proceedings of the International Conference on Ontologies, Databases, and Applications of Semantics (ODBASE)* (2013), pp. 501–518
28. T. Lukasiewicz, M.V. Martinez, G.I. Simari, Probabilistic preference logic networks, in *Proceedings of the European Conference on Artificial Intelligence (ECAI)* (IOS Press, Amsterdam, 2014), pp. 561–566
29. T. Lukasiewicz, M.V. Martinez, G.I. Simari, O. Tifrea-Marciuska, Ontology-based query answering with group preferences. ACM Trans. Internet Technol. **14**(4), 25 (2014)
30. T. Lukasiewicz, M.V. Martinez, G.I. Simari, O. Tifrea-Marciuska, Preference-based query answering in probabilistic Datalog+/– ontologies. J. Data Semant. **4**(2), 81–101 (2015)
31. T. Lukasiewicz, M.V. Martinez, D. Poole, G.I. Simari, Probabilistic models over weighted orderings: fixed-parameter tractable variable elimination, in *Proceedings of the International Conference on Principles of Knowledge Representation and Reasoning (KR)* (2016), pp. 494–504
32. S. Magliacane, A. Bozzon, E. Della Valle, Efficient execution of top-k SPARQL queries, *Proceedings of the International Semantic Web Conference (ISWC)* (2012), pp. 344–360
33. L. Moreau, The foundations for provenance on the Web. Found. Trends Web Sci. **2**(2/3), 99–241 (2010)
34. E. Ntoutsi, K. Stefanidis, K. Nørvåg, H. Kriegel, Fast group recommendations by applying user clustering, in *Proceedings of the International Conference on Conceptual Modelling (ER)* (Springer, Berlin, 2012), pp. 126–140
35. P.K. Pattanaik, *Voting and Collective Choice: Some Aspects of the Theory of Group Decision-Making* (Cambridge University Press, Cambridge, 1971)
36. M.S. Pini, F. Rossi, K.B. Venable, T. Walsh, Aggregating partially ordered preferences. Int. J. Log. Comput. **19**(3), 475–502 (2008)
37. F. Roberts, B. Tesman, *Applied Combinatorics* (CRC Press, Boca Raton, FL, 2009)
38. F. Rossi, K.B. Venable, T. Walsh, in *A Short Introduction to Preferences: Between Artificial Intelligence and Social Choice*. Synthesis Lectures on Artificial Intelligence and Machine Learning (Morgan & Claypool Publishers, San Rafael, CA, 2011)
39. W. Siberski, J.Z. Pan, U. Thaden, Querying the Semantic Web with preferences, in *Proceedings of the International Semantic Web Conference (ISWC)* (Springer, Berlin, 2006), pp. 612–624
40. B. Smith, G. Linden, Two decades of recommender systems at Amazon.com. IEEE Internet Comput. **21**(3), 12–18 (2017)
41. K. Stefanidis, G. Koutrika, E. Pitoura, A survey on representation, composition and application of preferences in database systems. ACM Trans. Database Syst. **36**(3), 19:1–19:45 (2011)
42. A.D. Taylor, *Social Choice and the Mathematics of Manipulation* (Cambridge University Press, Cambridge, 2005)

43. N. Wilson, Extending CP-nets with stronger conditional preference statements, in *Proceedings of the AAAI Conference on Artificial Intelligence (AAAI)*, vol. 4 (2004), pp. 735–741
44. M. Wooldridge, *An Introduction to Multiagent Systems* (Wiley, Hoboken, NJ, 2009)
45. X. Zhang, J. Chomicki, Preference queries over sets, in *Proceedings of the International Conference on Data Engineering (ICDE)* (2011), pp. 1019–1030

Chapter 3
Subjective Data: Model and Query Answering

User preferences have been incorporated in both traditional databases and ontology-based query answering mechanisms for some time now. The recent change in the way data is created and consumed in the Social Semantic Web has caused this aspect of query answering to receive more attention, since users play a central role in both knowledge engineering and knowledge consumption.

In this book, we focus on the problem of preference-based query answering in Datalog+/− ontologies assuming that querying users must rely on subjective reports of observing users to get a complete picture and make a decision. Real-world examples of this kind of situation arise when searching for products, such as smartphones or other mobile devices, booking a room in a hotel, or choosing a restaurant for dining. In all these cases, people use special-purpose applications or dedicated web sites, where they are prompted to provide some basic information in the search interface and then receive a list of answers (suggestions) to choose from, each associated with a set of subjective reports (often called reviews) written by other users to tell everyone about their experience.

The main problem with this setup, however, is that users are often overwhelmed and frustrated, because they cannot decide which reviews to focus on and which ones to ignore, since it is likely that, for instance, a very negative (or positive) review may have been produced on the basis of a feature that is completely irrelevant to the user who issued the query.

Consider the review in Fig. 3.1, where the reviewer provided a very negative result founded only on the reservation process and how a particular reservation was handled by the restaurant. A company or an individual looking for a place to hold a dinner for a large group may find this review relevant, and a tool suggesting options for that event should make the user take note of this report. On the other hand, the same tool, but being used by an individual looking for alternative options to dine on his own, could choose to not show this report unless the user is particularly interested in the reservation process, and could rather show reports related to the quality of food, or the value-for-money tradeoff.

© Springer International Publishing AG 2017
G.I. Simari et al., *Ontology-Based Data Access Leveraging Subjective Reports*,
SpringerBriefs in Computer Science, DOI 10.1007/978-3-319-65229-0_3

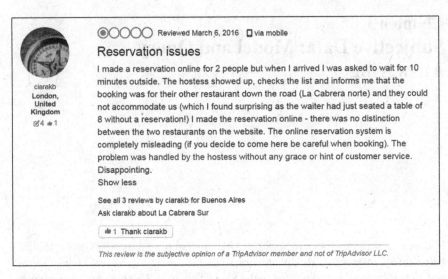

Fig. 3.1 An example of a very negative review in which the text focuses solely on the reservation process and how a particular reservation was handled by the restaurant. The user also did not fill out the value-based part of the review, though they did provide a location

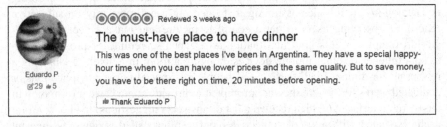

Fig. 3.2 An example of a very positive review focusing solely on happy-hour prices. The user did not fill out the value-based part of the review

In Fig. 3.2, the user gives a very positive report, but focuses only on the fact that happy-hour prices were available, which may or may not be relevant for other particular users.

In the rest of this chapter, we provide a formalization of the process of ranking answers with the help of the associated sets of subjective reports, and incorporating this ranking into preference-based query answering in Datalog+/− ontologies. Our approach is based on trust and relevance measures to select the best reports to focus on, given the user's initial preferences. The main goal is to obtain a user-tailored ranking over the set of query answers. We assume that each report contains scores for a list of *features*, its *author's preferences* among the features, as well as additional information (e.g., information on the reporter, such as age and nationality). These pieces of information in every report are then aggregated, along with the querying user's trust in each report, to a obtain a ranking of the query results relative to the preferences of the querying user (which are expressed over the same list of features that the reports refer to).

After introducing the underlying model (Sect. 3.1), we propose a basic approach to ranking query results (Sect. 3.2) where each atom is associated with an aggregate of the scores of all its reports—the aim is to present a ranking of the (top-k) atoms in the query result to the querying user. Every report's score is the average of the scores of each feature, weighted by: (1) the trust value for the feature score by the report, and (2) the relevance of the feature according the querying user's preferences (over the features). Also, to determine a report's score, the relevance of the report to the querying user's preferences is taken into account. Next, we present an alternative approach to ranking the query results, where we first select the most relevant reports for the querying user, adjust the scores by the trust measure, and compute a single score for each atom by combining the scores computed in the previous step, weighted by the relevance of the features. We describe algorithms for preference-based top-k (atomic) query answering in Datalog+/− ontologies under both rankings, proving that, under suitable assumptions, the two algorithms run in polynomial time in the data complexity. Finally, in Sect. 3.3, we propose and discuss a more general form of reports, which are associated with sets of atoms rather than single atoms.

3.1 A Logic-Based Data Model

Let *KB* be a Datalog+/− ontology, $a = p(c_1, \ldots, c_m)$ be a ground atom such that $KB \models a$, and $\mathcal{F} = (f_1, \ldots, f_n)$ be a tuple of *features* associated with the predicate p, each of which has a domain $dom(f_i) = [0, 1] \cup \{-\}$. We sometimes slightly abuse notation and use $\mathcal{F}$ to also denote the set of features $\{f_1, \ldots, f_n\}$.

A *report* for a is a triple $(E, \succ_P, I)$, where $E \in dom(f_1) \times \cdots \times dom(f_n)$, $\succ_P$ is an SPO over the elements of $\mathcal{F}$, and I is a set of pairs $(key, value)$. Intuitively, reports are evaluations of an entity of interest (atom a) provided by observers. In a report $(E, \succ_P, I)$, E specifies a "score" for each feature (with "−" meaning that no score has been provided), $\succ_P$ indicates the relative importance of the features to the report's observer, and I (called *information register*) contains general information about the report itself, and who provided it. Reports will be analyzed by a user, who has his own strict partial order, denoted $\succ_{P_U}$, over the set of features.

Example 3.1 Consider the restaurant domain from Example 1.3. There are many features that can be reviewed relative to a restaurant, such as location, atmosphere, service, menu variety, veggie-friendliness, parking, facilities for children, cleanliness, value for money, food (e.g., quality, preparation, presentation), and reservations (how they handled reservations in general). More fine-grained possibilities are: wine list, desserts, delivery, outdoor seating, and international friendliness (e.g., language). To keep the running example simple, we use the following six features for the predicate *restaurant*: $\mathcal{F} = (Atmosphere, Service, Reservation Process,$ *cleanliness, Value for Money, Food*); in the following, we abbreviate these features as *atm, serv, resP, clean, vfm,* and *food*, respectively.

An example of a report for *restaurant(laCabrera)* is $r_1 = (\langle 0.7, 0.4, 0.1, -0.2,$ $0.2\rangle, \succ_{P_1}, I_1)$, where $\succ_{P_1}$ is given by the graph in Fig. 3.3, and I_1 is a register with the fields *age_range, nationality, level of contribution* (how many reports the user contributed to the website; the higher the better), and *helpful votes* (the number of votes that the user received for proving helpful reports), with data $I_1.age$ being empty, $I_1.nationality = American$, $I_1.lvlContribution = 4$, and $I_1.hlpfVotes = 67$. This report corresponds to the one in Fig. 3.4.

An example of $\succ_{P_U}$ (i.e., the querying user's preferences over the features) is shown in Fig. 3.7.   ∎

The set of all reports available is denoted with *Reports*. In the following, we use *Reports(a)* to denote the set of all reports that are associated with a ground atom *a*. Given a tuple of features $\mathcal{F}$, we use $SPOs(\mathcal{F})$ to denote the set of all SPOs over $\mathcal{F}$.

3.1.1 Trust Measures over Reports

A user analyzing a set of reports may decide that certain opinions within a given report may be more trustworthy than others. For instance, returning to our running example, the score given for the *food* or *vfm* features of *restaurant(laCabrera)* in report r_1 might be considered more trustworthy than the ones given for *atm*, e.g., because the reviewer preference relation over the features shows the former to be among the most preferred features, while the latter is in the layer of the least preferred, cf. Fig. 3.3 (first report). Another example could be a user that is generally untrustworthy of reports on the feature *service*, because he has learned that many people are more critical than he is when evaluating that aspect of restaurants, or of reports on feature *food* by European users, because they tend to be very critical of any international food. Formally, a *trust measure* is any function $\tau: Reports \rightarrow [0, 1]^n$, where higher values correspond to more trust.

Note that trust measures do not depend on the querying user's own preferences over $\mathcal{F}$ (in $\succ_{P_U}$); rather, for each report $(E, \succ_P, I)$, they give a measure of trust to each of the *n* scores in *E* depending on *P* and *I*. The following is a simple example.

Example 3.2 Consider again our running example, and suppose that the user defines a trust measure τ, which assigns trust values to a report $r = (E, \succ_P, I)$, assigning more trust to those whose creator has "high" values corresponding to *level of contribution* or *helpful votes*, and it is defined as follows:

$$\tau(r) = \begin{cases} 0.25 \cdot \left(2^{-(rank(f_1, \succ_P)-1)}, \ldots, 2^{-(rank(f_n, \succ_P)-1)}\right) & \text{if } I.lvlContribution < 3 \\ & \text{or } I.hlpfVotes < 5; \\ \left(2^{-(rank(f_1, \succ_P)-1)}, \ldots, 2^{-(rank(f_n, \succ_P)-1)}\right) & \text{otherwise.} \end{cases}$$

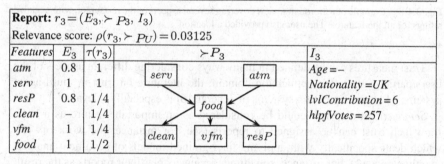

Report: $r_1 = (E_1, \succ P_1, I_1)$				
Relevance score: $\rho(r_1, \succ P_U) = 0.0156$				
Features	E_1	$\tau(r_1)$	$\succ P_1$	I_1
atm	0.7	1/32		*Age* = −
serv	0.4	1/16		*Nationality* = *USA*
resP	0.1	1/32		*lvlContribution* = 4
clean	−	1/32		*hlpfVotes* = 67
vfm	0.2	1/8		
food	0.2	1/4		

Report: $r_2 = (E_2, \succ P_2, I_2)$				
Relevance score: $\rho(r_2, \succ P_U) = 0.125$				
Features	E_2	$\tau(r_2)$	$\succ P_2$	I_2
atm	0.8	1/4		*Age* = −
serv	0.2	1/16		*Nationality* = *Argentina*
resP	−	1/32		*lvlContribution* = 2
clean	−	1/32		*hlpfVotes* = 1
vfm	0.2	1/4		
food	0.4	1/8		

Report: $r_3 = (E_3, \succ P_3, I_3)$				
Relevance score: $\rho(r_3, \succ P_U) = 0.03125$				
Features	E_3	$\tau(r_3)$	$\succ P_3$	I_3
atm	0.8	1		*Age* = −
serv	1	1		*Nationality* = *UK*
resP	0.8	1/4		*lvlContribution* = 6
clean	−	1/4		*hlpfVotes* = 257
vfm	1	1/4		
food	1	1/2		

Fig. 3.3 Reports used in Examples 3.5 and 3.6. These reports are for atom *restaurant*(*laCabrera*) and have been generated from Figs. 3.4, 3.5, and 3.6

For r_1 from Example 3.1, we get:

$$2^{-(rank(atm, \succ_{P_1})-1)} = 2^{-3} = 1/8,$$
$$2^{-(rank(clean, \succ_{P_1})-1)} = 2^{-3} = 1/8,$$
$$2^{-(rank(resP, \succ_{P_1})-1)} = 2^{-3} = 1/8,$$
$$2^{-(rank(serv, \succ_{P_1})-1)} = 2^{-2} = 1/4,$$
$$2^{-(rank(vfm, \succ_{P_1})-1)} = 2^{-1} = 1/2,$$
$$2^{-(rank(food, \succ_{P_1})-1)} = 2^{0} = 1.$$

Fig. 3.4 An example of a thorough review covering several aspects of the restaurant, with negative ratings for all the features. The user also provided a location

Trust measures can be defined in many ways considering different aspects of the users/participants in the application domain; the literature on trust in multi-agent systems [11] and reputation systems in general [6] is especially relevant.

Sentiment in a review could be considered as an important feature as a basis for which trust can be assigned to reports (see, for instance, the work of [7], which deals specifically with applying finer-grained models of sentiment analysis to online reviews). For instance, sometimes reviewers bias their reports as the result of extreme sentiments, and this results in a less rational, or much more subjective, opinion of the entity as a whole. As a user looking for references for a place to have dinner, one could prefer to ignore all such reviews charged with excessive emotional attitude. Figure 3.8 shows an example of this kind of review, along with the results of running its text on the IBM Watson Tone Analyzer.[1] There are many other popular tools available for sentiment analysis [3, 4, 8] (see [9, 10] for recent surveys on the topic), and they could be applied (on their own or in combination) as part of a trust measure in our framework.

---

[1] Available at: https://tone-analyzer-demo.mybluemix.net

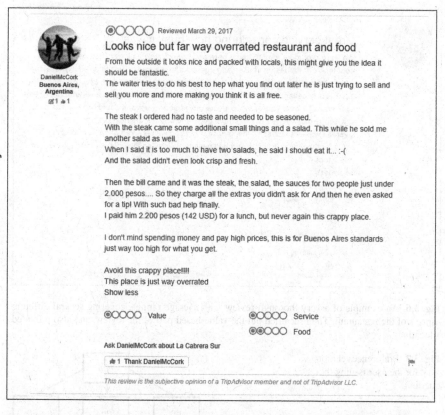

Fig. 3.5 An example of a thorough review, focusing mostly on service and food quality. The user also filled out the ratings with negative values for all the features, and also provided a location

Another important aspect to take into account when designing a trust measure is the possible presence of opinion *fraud* or *spam*, as well as *inconsistent* information when merging data from different sources. Chapter 4 includes a discussion of approaches to automatically detect such content (Sects. 4.2 and 4.4, respectively).

3.1.2   Relevance of Reports

The other aspect of importance that a user must consider when analyzing reports is how *relevant* they are to their own preferences. For instance, a report given by someone who has preferences that are completely opposite to those of the user should be considered less relevant than one given by someone whose preferences only differ in a trivial aspect. This is inherently different from the trust measure described above, as trust is computed without taking into account the preference relation given by the user issuing the query. Formally, a *relevance measure* is any

Fig. 3.6 An example of a very thorough review with average ratings, covering several different aspects of the restaurant. The user filled out the value-based part of the review, and also provided a location

Fig. 3.7 Preference relation $\succ_{P_U}$ for the user issuing the queries

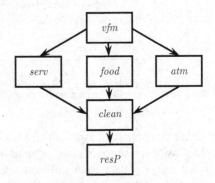

function $\rho \colon Reports \times SPOs(\mathcal{F}) \to [0, 1]$, where higher output values reflect higher relevance. Thus, a relevance measure takes as input a report $(E, \succ_P, I)$ and an SPO $\succ_{P'}$ and gives a measure of how relevant the report is relative to $\succ_{P'}$; this is determined on the basis of $\succ_P$ and $\succ_{P'}$, and can also take I into account.

Example 3.3 Consider again the running example, and suppose that the user assigns a relevance to a report $r = (E, \succ_P, I)$ according to the function

$$\rho(r, \succ_{P_U}) = 2^{-\sum_{f_i \in \mathcal{F}} |rank(f_i, \succ_P) - rank(f_i, \succ_{P_U})|}.$$

From Figs. 3.7 and 3.3, e.g., we have that $\rho(r_1, \succ_{P_U}) = 2^{-1 \cdot (2+1+0+1+1+1)} = 0.0156$.   ∎

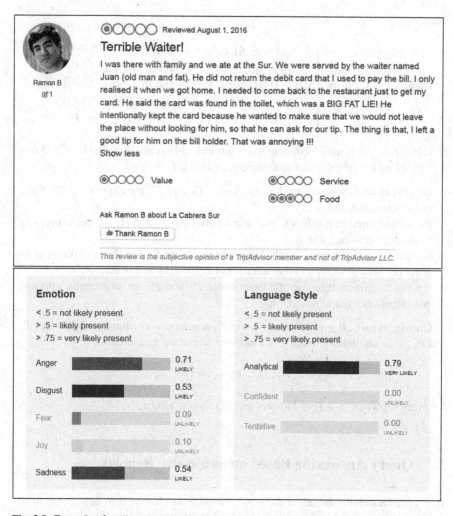

Fig. 3.8 Example of an "angry review" and the corresponding tone analysis results obtained by running the text through the IBM Watson Tone Analyzer; these results clearly indicate a high level of stress in the reviewer's tone, which could detract from the report's objectivity

Alternatively, a relevance measure comparing the SPO $\succ_P$ of a report $(E, \succ_P, I)$ with the user's SPO $\succ_{P_U}$ may be defined via measuring the similarity between two preference relations as follows.

Example 3.4 The relevance measure checks to what extent the two SPOs agree on the relative importance of the features in $\mathcal{F}$. Formally, let P_1 and P_2 be SPOs over $\mathcal{F}$. We define a measure of similarity of P_1 and P_2 as follows:

$$sim(P_1, P_2) = \frac{\sum_{1 \leq i < j \leq n} sim(f_i, f_j, P_1, P_2)}{n(n-1)/2},$$

where

$$sim(f_i, f_j, P_1, P_2) = \begin{cases} 1 & \text{if } (f_i, f_j) \in P_1 \cap P_2 \text{ or } (f_j, f_i) \in P_1 \cap P_2 \\ 1 & \text{if } (f_i, f_j) \notin P_1 \cup P_2 \text{ and } (f_j, f_i) \notin P_1 \cup P_2 \\ 0.5 & \text{if } ((f_i, f_j) \in P_1 \Delta P_2 \text{ and } (f_j, f_i) \notin P_1 \cup P_2) \text{ or} \\ & \quad ((f_j, f_i) \in P_1 \Delta P_2 \text{ and } (f_i, f_j) \notin P_1 \cup P_2) \\ 0 & \text{if } (f_i, f_j) \in P_1 \cup P_2 \text{ and } (f_j, f_i) \in P_1 \cup P_2. \end{cases}$$

Note that here, Δ is used to denote the symmetric difference (i.e., $A \Delta B = A \cup B - A \cap B$). More specifically, in the definition of $sim(f_i, f_j, P_1, P_2)$,

- the first condition refers to the case where P_1 and P_2 are expressing the same order between f_i and f_j,
- the second condition refers to the case where both P_1 and P_2 are not expressing any order between f_i and f_j,
- the third condition refers to the case where one of P_1 and P_2 is expressing an order between f_i and f_j and the other is not expressing any order,
- the last condition refers to the case where P_1 and P_2 are expressing opposite orders between f_i and f_j.

Clearly, $sim(P_1, P_2)$ is 1, when P_1 and P_2 agree on everything, and 0, when P_1 and P_2 agree on nothing. Finally, we define a relevance measure:

$$\rho((E, \succ_P, I), \succ_{P'}) = sim(\succ_P, \succ_{P'})$$

for every report $(E, \succ_P, I) \in Reports$ and SPO $\succ_{P'} \in SPOs(\mathcal{F})$.   ∎

3.2 Query Answering Based on Subjective Reports

To produce a ranking based on the basic components presented in Sect. 3, we must first develop a way to combine them in a principled manner. We first consider queries of the form $Q(\mathbf{X}) = p(\mathbf{X})$, called *atomic queries*, that is, queries with a single atom and no existential variables. The answers to an atomic query $Q(\mathbf{X}) = p(\mathbf{X})$ over *KB* in *atom form* are defined as $\{p(t) \mid t \in ans(Q(\mathbf{X}), KB)\}$; we still use $ans(Q(\mathbf{X}), KB)$ to denote the set of answers in atom form. At the end of this section, we will discuss how our approach can indeed be applied also to a more general class of queries called *simple*. The problem that we address is the following. The user is given a Datalog+/− ontology *KB* and has an atomic query $Q(\mathbf{X})$ of interest. The user also supplies an SPO $\succ_{P_U}$ over the set of features $\mathcal{F}$. Recall that in our setting, each ground atom b such that $KB \models b$ is associated with a (possibly empty) set of reports. As we consider atomic queries, each ground atom $a \in ans(Q(\mathbf{X}), KB)$ is an atom entailed by *KB* and thus is associated with a set of reports $Reports(a)$. Furthermore, recall that each report $r \in Reports(a)$ is associated with a trust score

$\tau(r)$. We want to rank the ground atoms in $Ans(Q(\mathbf{X}), KB)$, that is, we want to obtain a set $\{\langle a_i, score_i \rangle \mid a_i \in ans(Q(\mathbf{X}), KB)\}$ where $score_i$ for ground atom a_i takes into account the set of reports $Reports(a_i)$ associated with a_i, the trust score $\tau(r)$ associated with each report $r \in Reports(a_i)$, and the SPO $\succ_{P_U}$ over $\mathcal{F}$ provided by the user issuing the query.

3.2.1 A Basic Approach

A first approach to solving this problem is Algorithm RepRank-Basic (Fig. 3.9). A score for each atom is computed as the average of the scores of the reports associated with the atom, where the score of a report $r = (E, \succ_P, I)$ is computed as follows: (1) we first compute the average of the scores $E[i]$ weighted by the trust value for $E[i]$ and a value measuring how important feature f_i is for the user issuing the query (this value is given by $rank(f_i, \succ_{P_U})$); (2) then, we multiply the value computed in the previous step by $\rho(r, \succ_{P_U})$, which gives a measure of how relevant r is relative to $\succ_{P_U}$. The following is an example of how Algorithm RepRank-Basic works.

Algorithm RepRank-Basic($KB, Q(\mathbf{X}), \mathcal{F}, \succ_{P_U}, \tau, \rho, Reports, k$)
Input: Datalog+/– ontology KB,
          atomic query $Q(\mathbf{X})$,
          set of features $\mathcal{F} = \{f_1, \ldots, f_n\}$,
          preference relation of the querying user $\succ_{P_U}$,
          trust measure τ,
          relevance measure ρ,
          set of reports $Reports$,
          integer $k \geqslant 1$.
Output: Top-k answers to Q.

1. $RankedAns := \emptyset$;
2. for each atom a in $ans(Q(\mathbf{X}), KB)$ do begin
3. $score := 0$;
4.   for each report $r = (E, \succ_P, I)$ in $Reports(a)$ do begin
5. $trustMeasures := \tau(r)$;
6. $score := score + \rho(r, \succ_{P_U}) \cdot \frac{1}{n'} \cdot \sum_{i=1}^{n'} E[i] \cdot trustMeasures[i] \cdot \frac{1}{rank(f_i, \succ_{P_U})}$;
      // n' is the number of non-blank features in the report
7.   end;
8. $score := score / |Reports(a)|$;
9. $RankedAns := RankedAns \cup \{\langle a, score \rangle\}$;
10. end;
11. return top-k atoms in $RankedAns$.

Fig. 3.9 A first algorithm for computing the top-k answers to an atomic query Q according to a given set of user preferences and reports on answers to Q

Example 3.5 Consider again the setup from the running example, where we have the Datalog+/− ontology from Example 1.3, with a different database instance:

$$D = \{\; food(bifeDeChorizo),\quad food(parrillada),\quad foodType(meat),$$
$$type(bifeDeChorizo, meat),\quad type(parrillada, meat),$$
$$neighborhood(Palermo),$$
$$locatedIn(donJulio, Palermo),\quad locatedIn(laCabrera, Palermo),$$
$$serves(laCabrera, bifeDeChorizo),\quad serves(donJulio, parrillada)\}.$$

and the addition of the following two TGDs about neighborhoods:

$$r_6 : business(B) \to \exists N\; neighborhood(N) \wedge locatedIn(B, N),$$
$$r_7 : neighborhood(N) \to \exists C\; city(C) \wedge locatedIn(N, C).$$

Consider the set *Reports* of the reports depicted in Figs. 3.3 and 3.10, the SPO $\succ_{P_U}$ from Fig. 3.7, the trust measure τ defined in Example 3.2, and the relevance measure ρ introduced in Example 3.3. Finally, let $Q(X) = restaurant(X) \wedge locatedIn(X, buenosAires)$.

Algorithm RepRank-Basic iterates through the set of answers (in atom form) to the query, which in this case consists of:

$$\{restaurant(laCabrera), restaurant(donJulio)\}.$$

For the atom *restaurant(laCabrera)*, the algorithm iterates through the set of corresponding reports, which is $Reports(restaurant(laCabrera) = \{r_1, r_2, r_3\}$, and maintains the accumulated score after processing each report. For r_1, the score is computed as (cf. line 6 in Fig. 3.9):

$$0.0.00156 \cdot \frac{1}{5} \cdot \left(\frac{0.7 \cdot 0.03125}{2} + \frac{0.4 \cdot 0.0625}{2} + \frac{0.1 \cdot 0.03125}{4} + \right.$$
$$\left. \frac{0.2 \cdot 0.125}{1} + \frac{0.2 \cdot 0.25}{2} \right) = 0.00002314.$$

The score for *restaurant(laCabrera)* after processing the three reports is around 0.004916. Analogously, assuming $Reports(restaurant(donJulio)) = \{r_4, r_5, r_6\}$, the score for *restaurant(donJulio)* is 0.003215. Therefore, the top-2 answer to Q is:

$$\langle restaurant(laCabrera), restaurant(donJulio) \rangle.$$

∎

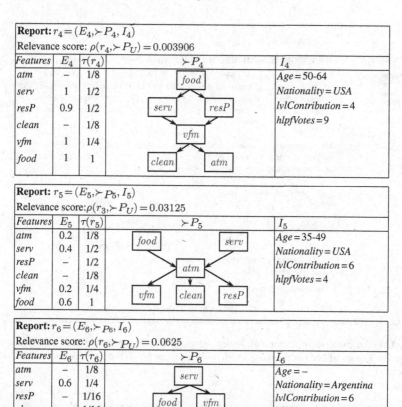

Fig. 3.10 Reports used in Examples 3.5 and 3.6. These reports are for atom *restaurant(donJulio)* and have been generated from Figs. 3.11, 3.12, and 3.13

The following result states the time complexity of Algorithm RepRank-Basic. As long as both query answering and computing the trust and relevance measures can be done in polynomial time, RepRank-Basic also runs in polynomial time.

Proposition 3.1 *The worst-case running time of Algorithm* RepRank-Basic *is:*

$$O\Big(m * \log m + (n + | \succ_{P_U} |) + m * Reports_{max} * (f_\tau + f_\rho + n) + f_{ans(Q(\mathbf{X}),KB)}\Big),$$

where $m = |ans(Q(\mathbf{X}), KB)|$, $Reports_{max} = \max\{|Reports(a)| : a \in ans(Q(\mathbf{X}),$ *KB)*$\}$, f_τ *(resp.,* f_ρ*) is the worst-case time complexity of* τ *(resp.,* ρ*), and* $f_{ans(Q(\mathbf{X}),KB)}$ *is the data complexity of computing* $ans(Q(\mathbf{X}), KB)$.

In the next section, we explore an alternative approach for applying the trust and relevance measures to top-k query answering.

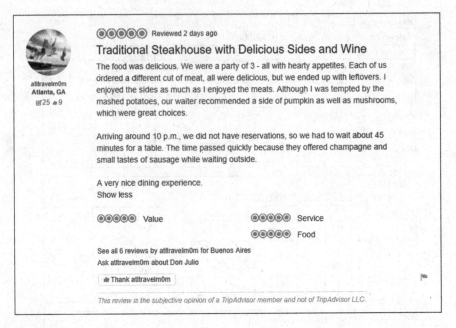

Fig. 3.11 An example of a thorough review, focusing mostly on food quality and free extras provided. The user filled out the ratings with positive values for all the features, and also provided a location

3.2.2 Leveraging Trust and Relevance to a Greater Extent

A more complex approach consists of using the trust and relevance scores provided by the respective measures in a more fine-grained manner. One way of doing this is via the following steps (more details on each of them are given shortly):

1. Keep only those reports that are most relevant to the user issuing the query, i.e., those reports that are relevant enough to $\succ_{P_U}$ according to a relevance measure ρ;
2. consider the most relevant reports obtained in the previous step and use the trust measure given by the user to produce scores adjusted by the trust measure; and
3. for each atom, compute a single score by combining the scores computed in the previous step with $\succ_{P_U}$.

The first step can simply be carried out by checking, for each report r, if $\rho(r, \succ_{P_U})$ is above a certain given threshold. One way of doing the second step is described in Algorithm SummarizeReports (Fig. 3.14), which takes a trust measure τ, a set of reports *Reports* (for a certain atom), and a function *collFunc*. The algorithm processes each report in the input sets by building a histogram of the average score over all reports per range of trust values (as collected in a bucket); for each report, the algorithm applies the trust measure to update each feature's histogram. Once all the reports are processed, the last step is to collapse the histograms into a

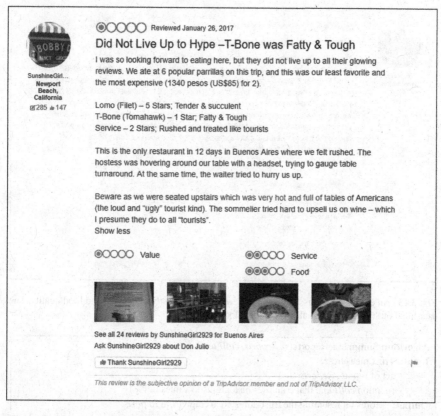

Fig. 3.12 An example of a thorough review, focusing mostly on food quality and atmosphere. The user also filled out the ratings and provided a location

single value—this is done by applying the *collFunc* function, which could simply be defined as the computation of a weighted average for each feature. This single value is finally used to produce the output, which is a tuple of n scores.

Example 3.6 Consider again the setup from Example 3.5. Suppose that we want to keep only those reports for which the relevance score is above 0.002 (as per the first step of our more complex approach). Recall that the answers to Q are {*restaurant(laCabrera)*, *restaurant(donJulio)*} and there are six associated reports; in this case, we trivially keep all records. Algorithm SummarizeReports has *Reports* = {r_1, r_2, r_3} when called for *restaurant(laCabrera)*. The histograms built during this call are as follows:

- *atm*: value 0.7 in bucket [0, 0.1), value 0.8 in bucket [0.2, 0.3), and value 0.8 in bucket [0.9, 1];
- *serv*: value 0.3 in bucket [0, 0.1) and value 1 in bucket [0.9, 1];
- *resP*: value 0.1 in bucket [0, 0.1) and value 0.8 in bucket [0.2, 0.3);
- *clean*: all the buckets are empty.

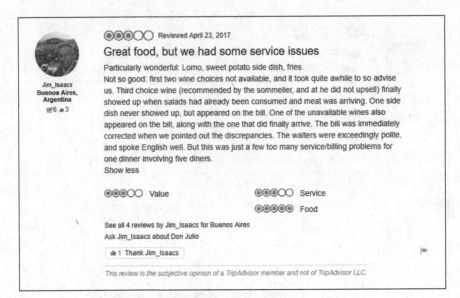

Fig. 3.13 An example of a less thorough review, focusing mostly on service and food quality. The user filled out the ratings and also provided a location

---

Algorithm SummarizeReports(τ, *Reports*, *collFunc*)

Input: Trust measure τ,

        set of reports *Reports*, and

        function *collFunc* that collapses histograms to values in $[0, 1]$.

Output: Scores representing the trust-adjusted average from *Reports*.

1. Initialize *hists*: n-array of empty mappings with keys: $\{[0, 0.1), [0.1, 0.2), \ldots, [0.9, 1]\}$ and values of type $[0, 1]$, where $n = |\mathcal{F}|$ (we use values $1, \ldots, 10$ to denote the keys).
2. Initialize array *bucketCounts* of size $n \times 10$ with value 0 in all positions;
3. for each report $r = (E, \succ_P, I) \in Reports$ do begin;
4.   *trustMeasures*:$= \tau(r)$;
5.   for $i = 1$ to n do begin
6.   let b be the key for *hists*$[i]$ under which *trustMeasures*$[i]$ falls;
7.   *hists*$[i](b)$:$= \frac{hist[i](b) * bucketCounts[i][b] + E[i]}{bucketCounts[i][b] + 1}$;
8.   *bucketCounts*$[i][b]$++;
9.   end;
10. end;
11. Initialize *Res* as an n-array;
12. for $i = 1$ to n do
13.   *Res*$[i]$:$= collFunc(hists[i])$;
14. return *Res*.

Fig. 3.14 This algorithm takes a set of reports for a single entity and computes an n-array of scores obtained by combining the reports with their trust measure

- *vfm*: value 0.2 in bucket $[0.1, 0.2)$ and value 0.6 in bucket $[0.2, 0.3)$.
- *food*: value 0.4 in bucket $[0.1, 0.2)$, value 0.2 in bucket $[0.2, 0.3)$, and value 1 in bucket $[0.5, 0.6)$.

Assuming that function *collFunc* disregards the values in the buckets corresponding to the two lowest trust values (if more than one bucket is non-empty, otherwise the largest value remains), and takes the average of the rest, we have the following result tuple as the output of SummarizeReports: $(0.8, 1, 0.8, -, 0.3, 0.1)$. Analogously, we have the tuple $(0.2, 0.65, 0.9, -, 0.6, 0.9)$ for *restaurant(donJulio)* after calling SummarizeReports with *Reports* = $\{r_4, r_5, r_6\}$. ∎

The following proposition states the time complexity of Algorithm Summarize-Reports. As long as the trust measure and the *collFunc* function can be computed in polynomial time, Algorithm SummarizeReports can be done in polynomial time as well.

Proposition 3.2 *The worst-case running time of Algorithm* SummarizeReports *is*

$$O(|Reports| * (f_\tau + n) + n * f_{collFunc}),$$

where f_τ *(resp.,* $f_{collFunc}$*) is the worst-case running time of* τ *(resp., collFunc).*

The following example explores a few different ways in which function *collFunc* used in Algorithm SummarizeReports might be defined.

Example 3.7 One way of computing *collFunc* is shown in Example 3.6, but there can be other reasonable ways of collapsing the histogram for a feature into a single value. For instance, *collFunc* might compute the average across all buckets ignoring the trust measure so that no distinction is made among buckets, i.e.:

$$collFunc(hists[i]) = \frac{\sum_{b=1}^{10} hists[i](b)}{10}.$$

Alternatively, the trust measure might be taken into account by giving a weight w_b to each bucket b (e.g., the weights might be set in such a way that buckets corresponding to higher trust scores have a higher weight, that is, $weight_i < weight_j$ for $i < j$). In this case, the histogram might be collapsed as follows:

$$collFunc(hists[i]) = \frac{\sum_{b=1}^{10} w_b * hists[i](b)}{10}.$$

We may also want to apply the above strategies but ignoring the first k buckets (for which the trust score is lower). Function *collFunc* can also be extended so that the number of elements associated with a bucket is taken into account. ∎

Thus, the second step discussed above gives n scores (adjusted by the trust measure) for each ground atom. Recall that the third (and last) step of the approach adopted in this section is to compute a score for each atom by combining the scores

---

Algorithm RepRank-Hist($KB, Q(\mathbf{X}), \mathcal{F}, \succ_{P_U}, \tau, \rho, relThresh, collFunc, Reports, k$)

Input: Datalog+/– ontology KB,

   atomic query $Q(\mathbf{X})$,

   set of features $\mathcal{F} = \{f_1, \ldots, f_n\}$,

   user preferences $\succ_{P_U}$,

   trust measure τ,

   relevance measure ρ,

   $relThresh \in [0, 1]$,

   function $collFunc$ that collapses histograms to values in $[0, 1]$,

   set of reports $Reports$,

   integer $k \geqslant 1$.

Output: Top-k answers to Q.

  1. $RankedAns := \emptyset$;
  2. for each atom a in $ans(Q(\mathbf{X}), KB)$ do begin
  3. $relReps :=$ select all $r \in Reports(a)$ with $\rho(r, \succ_{P_U}) \geqslant relThresh$;
  4. $scores :=$ SummarizeReports($\tau, relReps, collFunc$);
  5. $finalScore := \sum_{i=1}^{n} \frac{scores[i]}{rank(f_i, \succ_{P_U})}$;
  6. $RankedAns := RankedAns \cup \{\langle a, finalScore \rangle\}$;
  7. end;
  8. return top-k atoms in $RankedAns$.

Fig. 3.15 Algorithm for computing the top-k answers to an atomic query Q according to a given set of user preferences and reports on answers to Q

computed in the previous step with $\succ_{P_U}$. One simple way of doing this is to compute the weighted average of such scores where the weight of the i-th score is the inverse of the rank of feature f_i in $\succ_{P_U}$.

Algorithm RepRank-Hist (Fig. 3.15) is the complete algorithm that combines the three steps discussed thus far. The following continues the running example to show the result of applying this algorithm.

Example 3.8 Let us adopt once again the setup from Example 3.5, but this time applying Algorithm RepRank-Hist. Suppose $collFunc$ is the one discussed in Example 3.6 and thus Algorithm SummarizeReports returns the scores

$(0.8, 1, 0.8, -, 0.3, 0.1)$ for $restaurant(laCabrera)$ and
$(0.2, 0.65, 0.9, -, 0.6, 0.9)$ for $restaurant(donJulio)$.

Algorithm RepRank-Hist computes a score for each atom by performing a weighted sum of the scores in these tuples, which results in:

$$\langle restaurant(laCabrera), 1.5 \rangle, \quad \langle restaurant(donJulio), 1.7 \rangle.$$

Therefore, the top-2 answer to Q is:

$$\langle restaurant(donJulio), restaurant(laCabrera) \rangle,$$

which arrives at a different order than the one obtained in Example 3.5. ∎

Note that the results from the two algorithms are not necessarily the same, since they each use the relevance and trust score differently—the more fine-grained approach adopted by Algorithm RepRank-Hist allows it to selectively use both kinds of values to generate a more informed result.

Proposition 3.3 *The worst-case running time of Algorithm* RepRank-Hist *is:*

$$O\Big(m * \log m + (n + | \succ_{P_U} |) \Big) +$$

$$O\Big(m * \big(Reports_{max} * f_\rho + f_{sum} + n \big) + f_{ans(Q(\mathbf{X}),KB)} \Big),$$

where $m = |ans(Q(\mathbf{X}), KB)|$, $Reports_{max} = \max\{|Reports(a)| : a \in ans(Q(\mathbf{X}),$ $KB)\}$, f_ρ *is the worst-case time complexity of* ρ, f_{sum} *is the worst-case time complexity of Algorithm* SummarizeReports *as per Proposition 3.2, and* $f_{ans(Q(\mathbf{X}),KB)}$ *is the data complexity of computing* $ans(Q(\mathbf{X}), KB)$.

As a corollary to Propositions 3.1 and 3.3, we have the following result.

Theorem 3.1 *If the input ontology belongs to the guarded fragment of Datalog+/−, then Algorithms* RepRank-Basic *and* RepRank-Hist *run in polynomial time in the data complexity.*

Thus far, we have considered atomic queries. As each ground atom a such that $KB \models a$ is associated with a set of reports, and every ground atom b in $ans(Q(\mathbf{X}), KB)$ is such that $KB \models b$, reports can be associated with query answers in a natural way. We now introduce a class of queries that is more general than the class of atomic queries for which the same property holds. A *simple query* is a conjunctive query $Q(\mathbf{X}) = \exists \mathbf{Y}\, \Phi(\mathbf{X}, \mathbf{Y})$ where $\Phi(\mathbf{X}, \mathbf{Y})$ contains exactly one atom of the form $p(\mathbf{X})$, called *distinguished* atom (i.e., an atom whose variables are the query's free variables). For instance, $Q(X) = restaurant(X) \wedge locatedIn(X, Y)$ is a simple query, where $restaurant(X)$ is the distinguished atom. The answers to a simple query $Q(\mathbf{X})$ over KB in *atom form* are defined as $\{p(t) \mid t \in ans(Q(\mathbf{X}), KB)\}$, where the distinguished atom is of the form $p(\mathbf{X})$; we still use $ans(Q(\mathbf{X}), KB)$ to denote the set of answers in atom form. Clearly, for each atom a in $ans(Q(\mathbf{X}), KB)$, it is the case that $KB \models a$.

3.3 Towards More General Reports

In the previous section, we considered the setting where reports are associated with ground atoms a such that $KB \models a$. This setup is limited, since it does not allow us to express the fact that certain reports may apply to whole *sets* of atoms—this is necessary to model certain kinds of opinions often found in reviews, such as "all restaurants in *Palermo* are expensive and overrated". We now generalize the framework presented in Sects. 3 and 3.2.1 to contemplate this kind of reports.

A *generalized report* (g-report, for short) is a pair $gr = (r, Q(\mathbf{X}))$, where r is a report, and $Q(\mathbf{X})$ is a simple query, called the *descriptor* of gr. We denote with *g-Reports* the universe of g-reports. Intuitively, given an ontology KB, a g-report $(r, Q(\mathbf{X}))$ is used to associate report r with every atom a in $ans(Q(\mathbf{X}), KB)$—recall that $KB \models a$, and thus general reports allow us to assign a report to a set of atoms entailed by KB. Clearly, a report for a ground atom a as defined in Sect. 3 is a special case of a g-report in which the only answer to the descriptor is a.

Example 3.9 Consider our running example from the restaurant domain and suppose that we want to associate a certain report r with all the restaurants located in the neighbourhood of *Palermo* in Buenos Aires. This can be expressed with a g-report:

$$(r, Q(X)), \text{ with } Q(X) = \underline{restaurant(X)} \wedge locatedIn(X, palermo),$$

where *restaurant(X)* is the distinguished atom.   ∎

Intuitively, a g-report $gr = (r, Q(\mathbf{X}))$ is a report associated with a set of atoms, i.e., the set of atoms in $ans(Q(\mathbf{X}), KB)$. A simple way of handling this generalization would be to associate report r with every atom in this set. Note that, as in the non-generalized case, it could happen that two or more g-reports assign two distinct reports to the same ground atom. For instance, consider the g-report from Example 3.9 and another g-report of the form $(r', Q'(X))$, with:

$$Q'(X) = \underline{restaurant(X)} \wedge serves(parrillada) \wedge locatedIn(X, buenosAires),$$

expressing that r' applies to all restaurants in the city of Buenos Aires—in our running example, we would associate r with atoms *restaurant(laCabrera)* and *restaurant(donJulio)*, and r' only to *restaurant(donJulio)*.

In the approach just described, the reports coming from different g-reports are treated in the same way—they all have the same impact on the common atoms. Alternatively, we could exploit the ontology structure in order to determine when a g-report is in some sense *more specific* than another and to take such a relationship into account (e.g., more specific g-reports should have a greater impact when computing the ranking over atoms). We consider this kind of scenario in the following: we study two kinds of structures that can be leveraged from knowledge contained in the ontology. The first is based on the notion of *hierarchies*, which are useful in capturing the influence of reports in "is-a" type relationships. As an example, suppose we have the concept of *parrilla*, which is a kind of restaurant (typical Argentine denomination for a restaurant that serves grilled meat). Given a query requesting a ranking over restaurants in Palermo, a report for *all parrillas in Palermo* should have a higher impact on the ranking than a report for all restaurant in Buenos Aires—in particular, the latter might be ignored altogether, since it could be considered too general.

The second kind of structure is based on identifying subset relationships between the atoms associated with the descriptors in g-reports. For instance, a report for all restaurants in Palermo is more general than a report for all restaurants in Buenos Aires. We now define a partial order among reports based on these notions; we first define hierarchical TGDs as follows. A set of linear TGDs Σ_T is *hierarchical* iff for every $p(\mathbf{X}) \rightarrow \exists \mathbf{Y} q(\mathbf{X}, \mathbf{Y}) \in \Sigma_T$ we have that *features*$(p) \subseteq$ *features*(q) and there does not exist a database D over $\mathcal{R}$ and TGD in Σ_T of the form $p'(\mathbf{X}) \rightarrow \exists \mathbf{Y} r(\mathbf{X}, \mathbf{Y})$ such that $p(\mathbf{X})$ and $p'(\mathbf{X})$ share ground instances relative to D.

In the rest of this section, we assume that all ontologies contain a (possibly empty) subset of hierarchical TGDs. Furthermore, given an ontology $KB = (D, \Sigma)$, where $\Sigma_H \subseteq \Sigma$ is a set of hierarchical TGDs, and two ground atoms a and b, we say that a *is-a* b iff *chase*$(\{a\}, \Sigma_H) \models b$. For instance, in Example 1.3, the set $\{r_1\} \subseteq \Sigma$ is a hierarchical (singleton, in this case) set of TGDs assuming that the conditions over the features hold (clearly, for the concept *business*, the feature *food* is not applicable, since business in general could provide services and/or products not necessarily related to gastronomy). On the other hand, if we add to the ontology the concept of *parrilla* (described above), then a TGD:

$$r_8 : parrilla(X) \rightarrow restaurant(Y)$$

is a hierarchical TGD.

Given tuples of features $\mathcal{F}$ and $\mathcal{F}'$ such that $\mathcal{F} \subseteq \mathcal{F}'$ and vectors E and E' over the domains of $\mathcal{F}$ and $\mathcal{F}'$, respectively, we say that E' is a *particularization* of E, denoted $E' = part(E)$ iff $E'[f] = E[f]$, if $f \in F \cap F'$, and $E'[f] = -$, otherwise. Let $KB = (D, \Sigma)$ be a Datalog+/− ontology, a be a ground atom such that $KB \models a$, and $gr = (r, Q(\mathbf{X}))$ be a g-report with $r = (E, \succ_P, I)$. If there exists a ground atom $b \in Ans(Q(\mathbf{X}), KB)$ such that a *is-a* b, then we say that g-report $gr' = ((E', \succ_P, I), a)$, with $E' = part(E)$, is a *specialization* of gr for a. Clearly, a g-report is always a specialization of itself for every atom in the answers to its descriptor.

Example 3.10 Let $\mathcal{F}_1$ be the set of features for predicate *restaurant* presented in Example 3.1, and let $\mathcal{F}_2 = (Atmosphere, Service, Reservation Process, cleanliness, Value for Money, food, grill)$ be the set of features for predicate *parrilla*, where *grill* evaluates the quality of the grilling process and resulting food.

Let $gr = (r_1, Q(X))$ be a g-report, where r_1 is the report from Fig. 3.3 and

$$Q(X) = restaurant(X) \wedge locatedIn(X, Palermo).$$

Consider $a = restaurant(laCabrera)$, $b = parrilla(laCabrera)$, and TGD r_8; clearly, we have that $b \in Ans(Q(X), KB)$ and b *is-a* a. Therefore, a specialization of gr for b is $gr' = ((E', \succ_{P_1}, I_1), b)$, where $E' = \langle 0.7, 0.4, 0.1, -, 0.2, 0.2, - \rangle$; the last "−" indicates that there is no valuation for the specific feature *grill* from r_1. ∎

Given g-reports $gr_1 = (r_1, Q_1(\mathbf{X}_1))$ and $gr_2 = (r_2, Q_2(\mathbf{X}_2))$, we say that gr_1 is *more general than* gr_2, denoted $gr_2 \sqsubseteq gr_1$, iff either (i) $Ans(Q_2(\mathbf{X}_2), KB) \subseteq Ans(Q_1(\mathbf{X}_1), KB)$; or (ii) for each $a \in Ans(Q_2(\mathbf{X}_2), KB)$, there exists some $b \in Ans(Q_1(\mathbf{X}_1), KB)$ such that a is-a b. If $gr_1 \sqsubseteq gr_2$ and $gr_2 \sqsubseteq gr_1$, we say that gr_1 and gr_2 are *equivalent*, denoted $gr_1 \equiv gr_2$.

$$gr_1 = \left((r_1, \succ_{P_1}, I_1), \underline{restaurant}(X) \wedge locatedIn(X, buenosAires)\right)$$
$$gr_2 = \left((r_4, \succ_{P_1}, I_1), \underline{parrilla}(X) \wedge locatedIn(X, palermo)\right)$$

Fig. 3.16 A set of general reports (distinguished atoms in the descriptors are *underlined*)

Example 3.11 Consider the following example ontology:

$$\Sigma_T = \{\ r_1 : restaurant(R) \rightarrow business(R),$$
$$r_2 : parrilla(P) \rightarrow restaurant(X),$$
$$r_4 : locatedIn(X, Y) \wedge locatedIn(Y, Z) \rightarrow locatedIn(X, Z)\},$$

$$D = \{\ restaurant(donJulio),\quad parrilla(laCabrera),$$
$$locatedIn(donJulio, Palermo),\quad locatedIn(laCabrera, Palermo)\}.$$

For this example, assume that the reports r_1 and r_4 have been given for "*all restaurants in Buenos Aires*" and "*all parrillas in Palermo*", respectively, as described in Fig. 3.16. Consider the ontology in the running example, replacing the atom *restaurant(laCabrera)* with *parrilla(laCabrera)*. We then have that $gr_2 \sqsubseteq gr_1$, since

$$\{restaurant(laCabrera)\} \subseteq \{restaurant(laCabrera), restaurant(donJulio)\}.$$

∎

The "more general than" relationship between g-reports is useful for defining a partial order for the set of reports associated with a given ground atom. This partial order can be defined as follows:

$$gr_1 \sim gr_2 \text{ iff } gr_1 \equiv gr_2,$$

and

$$gr_1 \succ gr_2 \text{ iff } gr_1 \sqsubseteq gr_2.$$

Here, $a \sim b$ denotes the equivalence between a and b. This relationship can then be used to define weighting functions that allow us to assign importance to reports based on their generality. Such a weighting function can be defined as any function $\omega : g\text{-}Reports \rightarrow [0, 1]$ such that:

1. if $gr_1 \succ gr_2$, then $\omega(gr_1) > \omega(gr_2)$; and
2. if $gr_1 \sim gr_2$, then $\omega(gr_1) = \omega(gr_2)$.

For example, one possible weighting function is defined as $\omega(gr) = 2^{-rank(gr, \succ)+1}$.

Leveraging Multi-Dimensional Data Models Recently, there have been developments in multi-dimensional aspects of knowledge bases. In particular, the work in [1, 2] proposes a multi-dimensional data model based on Datalog$+/-$ and previous work on Hurtado-Mendelzon's model [5] for multi-dimensional data. These works allow to interpret data from different contexts (or dimensions), providing a logical representation of dimension hierarchies, dimensional constraints, and dimensional rules that connect the different levels in the model. This model can be leveraged in our framework using dimensions to structure the relationships that allow us to reason about general reports. For instance, in our running example, the ontology models a gastronomical domain—here, potential dimensions could structure the data about location, food, and business taxonomies. In this case, the sets of hierarchical TGDs do not need to be part of the application domain conceptualization, and can be used in a different domain modeling, for instance, other types of services.

References

1. L.E. Bertossi, M. Milani, The ontological multidimensional data model, in *Proceedings of the Alberto Mendelzon Workshop on Foundations of Data Management (AMW)* (2017)
2. L.E. Bertossi, M. Milani, The ontological multidimensional data model (extended abstract). CoRR, abs/1703.03524 (2017)
3. E. Cambria, Affective computing and sentiment analysis. IEEE Intell. Syst. **31**(2), 102–107 (2016)
4. J.B. Hollander, E. Graves, H. Renski, C. Foster-Karim, A. Wiley, D. Das, A (short) history of social media sentiment analysis, in *Urban Social Listening* (Springer, Berlin, 2016), pp. 15–25
5. C.A. Hurtado, A.O. Mendelzon, OLAP dimension constraints, in *Proceedings of the ACM SIGMOD-SIGACT-SIGART Symposium on Principles of Database Systems (PODS)* (ACM, New York, 2002), pp. 169–179
6. E. Koutrouli, A. Tsalgatidou, Reputation systems evaluation survey. ACM Comput. Surv. **48**(3), 35 (2016)
7. F.V. Ordenes, S. Ludwig, D. Grewal, K. de Ruyter, M. Wetzels, Analyzing online reviews through the lens of speech act theory: implications for consumer sentiment analysis. J. Consum. Res. (2016)
8. D.R. Recupero, V. Presutti, S. Consoli, A. Gangemi, A.G. Nuzzolese, Sentilo: frame-based sentiment analysis. Cogn. Comput. **7**(2), 211–225 (2015)
9. K. Schouten, F. Frasincar, Survey on aspect-level sentiment analysis. IEEE Trans. Knowl. Data Eng. **28**(3), 813–830 (2016)
10. A. Yadollahi, A.G. Shahraki, O.R. Zaiane, Current state of text sentiment analysis from opinion to emotion mining. ACM Comput. Surv. **50**(2), 25 (2017)
11. H. Yu, Z. Shen, C. Leung, C. Miao, V.R. Lesser, A survey of multi-agent trust management systems. IEEE Access **1**, 35–50 (2013)

Chapter 4
Related Research Lines

As we have already discussed in previous chapters, the work described in this book can be considered as a further development of the PrefDatalog+/− framework presented in [9] (cf. Chap. 2), where we develop algorithms to answer skyline queries, and their generalization to k-rank queries, over classical Datalog+/− ontologies. However, PrefDatalog+/− assumes that a model of the user's preferences is given at the time the query is issued. In the model and algorithms discussed in Chap. 3, we make no such assumption; instead, we assume that the user only provides some very basic information regarding their preferences over certain features, and that they have access to a set of reports provided by other users in the past. In a sense, this approach is akin to building an ad hoc model *on the fly* at query time and using it to provide a list of results that is ordered with respect to the querying user's preferences.

In the rest of this chapter, we will discuss several aspects that touch on reasoning with and managing online reviews produced by people. As we will see, there are several formidable hurdles that must be overcome in this data (and cognitive) engineering task.

4.1 Assessing the Impact of Reviews

It is a well-known fact that modern-day users often consult reviews available on different platforms when they are in the process of booking hotels, flights, and tour packages online. Some interesting statistics say, for instance, that 90% of consumers read online reviews before visiting a business, 88% trust online reviews as much as personal ones, 72% will take action only after reading a positive review, and 86% of people will hesitate to purchase from a business that has negative online reviews [18]—however, the impact of reviews has not been studied in a systematic, well-founded way. In [22], an early work from the tourism management

© Springer International Publishing AG 2017

G.I. Simari et al., *Ontology-Based Data Access Leveraging Subjective Reports*,
SpringerBriefs in Computer Science, DOI 10.1007/978-3-319-65229-0_4

literature, the authors carry out an experimental study of how different aspects of reviews impact consumers when considering hotels to book: review valence (positive/negative), hotel familiarity (how well-known they are), and reviewer expertise (how experienced they are). They found that online reviews causes hotels to enjoy more customer consideration; interestingly, even negative reviews have some positive effects, since they increase awareness (the effects were found to be stronger for lesser-known hotels). Finally, another interesting conclusion was that reviewer expertise has only a small positive effect on review impact.

Another interesting phenomenon, studied in [2], is that of reviews carried out by people who have not actually made a purchase. Though, it is not easy to conclusively determine that any given review does not have an associated purchase (because perhaps the reviewer is posting it on Amazon when they actually purchased from eBay, for instance), the authors point out that there is a non-trivial cluster of reviews—about 5%—that do not have an associated purchase on the site and are also significantly more negative, with specific aspects pointing to deceptive language (see below for a discussion on detecting this kind of reviews). Though, it is difficult to understand precisely why people engage in this sort of behavior, the authors propose three possibilities: (1) that loyal customers act as self-appointed brand managers; (2) that they are upset customers; or (3) that these customers are trying to enhance their social status.

Finally, another interesting line of work in this category is that of determining and predicting the *helpfulness* of reviews [23]—this is an important problem to solve, since popular products and services often have too many reviews for customers to read in full, and so they need to be ranked accordingly. The authors point out that previous work on this problem produced models that are not transferable, because they leveraged features apart from the review text itself. In contrast, their approach focuses solely on the text, and an experimental evaluation shows that performance is improved with respect to the alternative methods, the resulting models are more easily generalized, and they are highly interpretable. Figure 4.1 shows an example of how review helpfulness plays a role in Amazon's presentation of product reviews and ratings to its users.

4.2   Fraud and Spam Detection

As was briefly mentioned in the previous section, one of the main problems that needs to be solved when using online reports is that of automatically detecting spam, fraud, and general trustworthiness of the opinions being expressed.

To our knowledge, the first work on detecting spam in reviews is only a decade old [7]. This is somewhat surprising, given that spam in web sites and email messages has been the topic of research for a long time; perhaps, this is a reflection of how daunting the task is, given that this type of content is inherently subjective.

Fig. 4.1 Example of Amazon's efforts to present reviews based on helpfulness. The tooltip shown underneath the star rating of the product states that review helpfulness is taken into account in its calculation

The authors analyze 5.8M reviews of 6.7M products on amazon.com, written by 2.14M reviewers, and discovered that spam occurs quite prominently. The authors propose to classify spam reviews into three main categories:

- **Type 1: False opinions.** Reviews in this category are simply false—they can be either positive (designed to promote the product) or negative (aiming to damage the product's or company's reputation).
- **Type 2: Reviews on brands only.** They do not refer to products, but rather only to brands, manufacturers, or merchants. Although these reviews express opinions that may be valid and useful (as seen in the g-reports of Chap. 3), they can also be highly biased, such as reviews written by haters or blind fans.
- **Type 3: Non-reviews.** These reviews do not contain any opinions; though, they are generally innocuous, since they can be easily detected, they do bias automatic aggregators that compute, for instance, average ratings. The authors identify two sub-categories: (1) *advertisements* focusing on listing product features, which can be of the actual product, of competing products, or even references to other (competing) sellers; and (2) *other non-reviews*, which can be questions/answers, comments referring to other reviews, or random/unrelated text.

Figure 4.2 shows two reviews of restaurants on Yelp with several warning signs pointing to possible spam: the reviewers have no friends and no picture, this is their only review, and the opinions expressed are quite vague and can be seen as designed to hurt the restaurant's reputation—possibly spam of Type 1.

Reviews of Type 1 are clearly the most difficult to detect—even manually—given that carefully prepared false opinions are indistinguishable from valid ones.

Irina M.
Manhattan, NY
0 friends
1 review

★★★★★ 4/18/2017

Don Julio comes highly recommended. It's in all the travel guides. The service is terrific, the location quite nice. And the veggie parrillada was very tasty and well seasoned. But - you don't go to an Argentinian parrilla for veggies. We had the sirloin steak mariposa and - seriously- had better steak at the airport. We requested the steak rare (jugoso) and it was served medium well (a punto). It was underseasoned, a bit dry, and the meat itself lacked flavor - a rarity in Argentina. In summation, while the steak may be passable by the standards in the rest of the world (which might explain its high ranking on yelp); by Argentinian standards Don Julio is at best mediocre, and more accurately subpar. Anyone thinking of going there would be best served by saving half their money and enjoying any corner parrilla.

Was this review ...?

🏆 Useful 1   😄 Funny 1   😎 Cool

Kris P.
Austin, TX
0 friends
1 review

★★★★★ 1/6/2017

The service is fast and responsive for Buenos Aires and it's a nice enough atmosphere. The food and experience overall is just okay. It definitely has a "tourist trap" feel to it and the meat they serve reflects that. I'd try La Brigada or Don Julio before booking a rez here.

Was this review ...?

🏆 Useful   😄 Funny   😎 Cool

Fig. 4.2 Low quality (possibly spam by competitor) reviews of restaurants Don Julio (*top*) and La Cabrera (*bottom*) on Yelp

One phenomenon that tends to occur along with Type 1 reviews, however, is that of duplicate (or near duplicate) reviews, which can be either duplicate text from different user ids on the same product, duplicate text from the same user id but on different products, and duplicate text from different user ids on different products.

Later work by some of the same authors [8] considered further information based on user behavior in order to aid in the process of detecting review spam. In particular, certain products or product groups may be targeted by spammers to maximize the impact, and spam tends to deviate substantially from other reviews, particularly in their ratings. The resulting methods were evaluated also on Amazon data, and they outperformed the baseline method using just helpfulness votes.

The work of [14] also focused on behavioral patterns as a means to detect spam, proposing an unsupervised learning clustering model called the *Author Spamicity Model* (ASM) that works in the Bayesian setting—this is the first attempt at a general solution to this problem, since previous works were mostly developed ad hoc. The main author features considered are content similarity, maximum number of reviews in a day, reviewing burstiness (based on the difference between the first and last review posting dates), and ratio of first reviews (how many of the author's reviews are the first for the product). For reviews, the features considered are: duplicate/near duplicate reviews, extreme ratings, rating deviation, early time frame (similar to the ratio of first reviews feature for authors), and rating abuse (such as multiple ratings for the same product). A thorough empirical evaluation of the approach was carried out with very good results.

Another approach to detecting opinion spam was taken in [16], where the authors frame the task in terms of a psycholinguistic deception detection problem—essentially, deceptive statements are expected to be similar to lies where, for instance, increased negative emotions and psychological distancing are present. As an alternative, they also propose to view the problem as a genre identification one, where deceptive and truthful writings are sub-genres of imaginative and informative writings, respectively. Experimental results obtained via this approach reach nearly 90% accuracy. Another interesting result of their empirical evaluation is that human subjects perform quite poorly, almost at the level of random choice.

Stylometry has also been applied to this problem [4], where the main contribution was to use features from context free grammars instead of the more traditional shallow syntactic features. Experimental results from several datasets show that the approach works very well, reaching 91.2% accuracy.

In [1], the FraudEagle tool is presented. The novel aspect in this approach was the use of network properties in the detection of fraudulent content—networks represent how users and products are linked via reviews, which are labeled with signs (positive or negative). This tool is complementary to previous approaches based on text or behavior, and priors used in the computation of classifications can be seeded with such clues. Experimental results on both synthetic and real-world data show the viability of the approach.

Finally, in [19], the authors show a very simple technique for producing fake reviews, based on taking a truthful review as a template and replacing sentences with those from other reviews—such reviews are very difficult to detect, both by state-of-the-art methods as well as humans. The authors also present a framework to detect reviews that are machine-synthesized in this manner, based on measures of semantic coherence and flow smoothness.

For a recent survey on this topic, we refer the interested reader to [5], where the main approaches developed in the literature are analyzed and compared with respect to their strengths and weaknesses, and assessed in terms of their accuracy and results.

4.3 Review Summarization

Information overload is one of the problems that readers of online reviews must face when dealing with popular products and services; it is also a problem for merchants and providers, since keeping track of customers' opinions is also important for

Hotel details

Set in a downtown tower block, this contemporary hotel is a 12-minute walk from Club de Golf del Uruguay, a 14-minute walk from the nearest beach and 20 km from Carrasco International ... more

Review summary ⓘ   [Write a review]   [Add a photo]

Rooms · 2.8 ★★★★★
Guests liked the large rooms, though some said they were dated & cleanliness & maintenance could be improved · Some guests said the bathrooms were small & cleanliness could be improved

Location · 4.4 ★★★★★
Near the beach; shopping, restaurants & bars nearby

Service & facilities · 3.6 ★★★★★
Guests appreciated the friendly staff · Some guests said reception could be improved · Guests enjoyed the fitness center

View Google reviews

Fig. 4.3 Sample review summary offered by Google when searching for a hotel. Reviews are separated into ratings and opinions on three different features, giving pros and cons for each one when available

them. The task of *review summarization* aims at tackling this problem by producing summaries of the opinions expressed in reviews [6]. The main steps involved in such a task are: (1) mining the features that are the subjects of reviews; (2) identifying which parts of the reviews refer to which features, and the sentiment expressed in each case; and (3) producing a summary of the results. Figure 4.3 shows an actual example of how Google is currently incorporating this kind of summary in the information box on the right hand side of the screen when searching for hotels.

Relatively mature systems for review summarization have been available for several years. For instance, Review Spotlight [24] was developed as a Google Chrome extension that yields brief overviews of information contained in reviews in the form of adjective-noun pairs, such as *"best sushi"*, *"reasonable price"*, or *"long wait"*, with color coding to indicate the sentiment associated with each one (also allowing to access more detailed information by clicking on the pair). This system was one of the first to tackle the problem from the point of view of user interfaces.

Recent work has also addressed summarization of *micro-reviews* [11], which are very brief observations like *"Try the Beach Salad!"* and *"They have #The12thCan!!"*—FourSquare sends this kind of snippet to users when they approach a venue. The authors focus on the problem of providing a short list comprising the most helpful micro-reviews—this requires combining entity recognition, sentiment analysis, and summarization. A crowdsourced empirical evaluation of their resulting system showed that microsummaries are preferred over traditional summaries based on text extraction.

4.4 Fact Finding and Inconsistency Detection

The problems of *fact finding* and *inconsistency detection* are somewhat tangential to the main topic of this book, but adequately addressing them is quite important in systems that leverage user-provided reports—this is simply because it is often useful to consider information coming from multiple sources. Therefore, in this kind of data integration scenario, information sources need to be found in order to populate values for the different properties of objects. However, there is no guarantee that the information found is correct (for instance, an old phone number might still appear on some sites). Furthermore, conflicting data may also be present (both old and new phone numbers appear, depending on the size), or values for specific fields might be missing (for instance, opening hours might not be available).

The work of [25] tackles such problems by proposing the TruthFinder algorithm, which analyzes the relationship between web sites and the individual facts that they claim. Towards this end, a series of basic heuristics are used: (1) there is usually a single correct value for a property of an object; (2) correct values appear in multiple places, with possible slight variations; (3) incorrect values for a property of an object tend to be different/not similar; and (4) for a given domain, if a site provides many correct values for different property-object pairs, it will likely provide correct values for others as well. Based on these heuristics, the authors devise methods for computing fact confidence and web site trustworthiness, and carry out empirical evaluations showing that they obtain high accuracy in their results.

In [17], the authors address these problems by introducing a framework for incorporating *prior knowledge* into fact-finding algorithms, claiming that this can greatly help obtain higher-quality results. Prior knowledge can come in the form of common sense or specific facts.

Figure 4.4 shows an example of several inconsistencies and of incompleteness between the information available for restaurant *La Cabrera Norte* in Buenos Aires on TripAdvisor and Yelp:

- Business hours are available on one but not the other.
- The price ranges differ quite significantly.
- TripAdvisor has it listed as a lunch and dinner place, while Yelp says "good for dinner".
- TripAdvisor says that it has outdoor seating, while Yelp says it does not.
- Information on electronic payment options is not inconsistent, but has different level of specificity.
- Alcohol availability is also vaguely stated on TripAdvisor as "serves alcohol" and "wine and beer", while Yelp states "beer and wine only".

Finally, note that the location information is correct on both sites, but the name of the street appears in full form on TripAdvisor, and abbreviated on Yelp, the neighborhood on TripAdvisor is more precise (Palermo Soho vs. Palermo); even if the map markers seem to be in different places, this is the result of one map being rotated 90° with respect to the other.

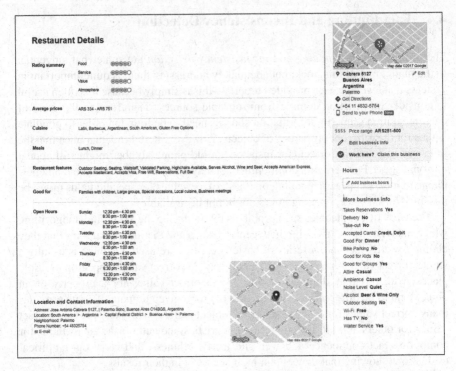

Fig. 4.4 Example of inconsistent information for restaurant *La Cabrera Norte* from TripAdvisor (*left*) and Yelp (*right*). The image is the result of composing the information available on each site for illustrative purposes

4.5   Improving Recommender Systems Leveraging Review Text

Recently, there has been an increasing interest in leveraging information that can be found in reviews to improve recommendations. For instance, in [3, 12, 20, 27], the focus is on the problem of leveraging the text in reviews to supplement missing information in rating scores, as well as to learn topics. Another similar approach was taken in [15], where user profiles are built based on users' review texts, and then leveraged to filter other reviews with "the eyes of the user". Another related approach, investigated in [26], proposes a novel model to contemplate the local context of words and add further depth to models of users, items, and reviews. Figure 4.5 shows an example of how Yelp makes use of the reviews available on the site to give recommendations to its users.

Best of Yelp Buenos Aires – Steakhouses

See More **Steakhouses in Buenos Aires**

Fig. 4.5 Recommendations provided by Yelp to users viewing the page for restaurant Don Julio

Surprisingly, review text was largely ignored in earlier approaches that make use of user feedback—however, empirical evaluations in these works show promising results for this kind of framework. A related approach leverages social relations with the same objective [10]; recent work combining the two, obtaining good results in practice, is proposed in [21].

Finally, another recent work leverages user-provided reviews to learn two kinds of relationships between products: substitutable (two different, but similar, phones) and complementary (such as phones and chargers) [13]. The method developed in that work is based on a topic modeling task to derive these relationships by means of supervised learning via link prediction, making use not only of review texts, but also other features outside the realm of user reports, such as product specifications and prices.

References

1. L. Akoglu, R. Chandy, C. Faloutsos, Opinion fraud detection in online reviews by network effects, in *International AAAI Conference on Weblogs and Social Media (ICWSM)* (2013)
2. E.T. Anderson, D.I. Simester, Reviews without a purchase: low ratings, loyal customers, and deception. J. Mark. Res. **51**(3), 249–269 (2014)
3. Y. Bao, H. Fang, J. Zhang, TopicMF: simultaneously exploiting ratings and reviews for recommendation, in *Proceedings of the AAAI Conference on Artificial Intelligence (AAAI)* (AAAI Press, Palo Alto, 2014), pp. 2–8
4. S. Feng, R. Banerjee, Y. Choi, Syntactic stylometry for deception detection, in *Proceedings of the Annual Meeting of the Association for Computational Linguistics (ACL): Short Papers–Volume 2* (Association for Computational Linguistics, Stroudsburg, 2012), pp. 171–175
5. A. Heydari, M. Tavakoli, N. Salim, Z. Heydari, Detection of review spam: a survey. Expert Syst. Appl. **42**(7), 3634–3642 (2015)
6. M. Hu, B. Liu, Mining and summarizing customer reviews, in *Proceedings of the International Conference on Knowledge Discovery and Data Mining (SIGKDD)* (ACM, New York, 2004), pp. 168–177

7. N. Jindal, B. Liu, Analyzing and detecting review spam, in *Proceedings of the IEEE International Conference on Data Mining (ICDM)* (IEEE, Washington, DC, 2007), pp. 547–552
8. E.P. Lim, V.A. Nguyen, N. Jindal, B. Liu, H.W. Lauw, Detecting product review spammers using rating behaviors, in *Proceedings of the ACM International Conference on Information and Knowledge Management (CIKM)* (ACM, New York, 2010), pp. 939–948
9. T. Lukasiewicz, M.V. Martinez, G.I. Simari, Preference-based query answering in Datalog+/− ontologies, in *Proceedings of the International Joint Conference on Artificial Intelligence (IJCAI)* (2013), pp. 1017–1023
10. H. Ma, H. Yang, M.R. Lyu, I. King, SoRec: social recommendation using probabilistic matrix factorization, in *Proceedings of the ACM Conference on Information and Knowledge Management (CIKM)* (ACM, New York, 2008), pp. 931–940
11. R. Mason, B. Gaska, B. Van Durme, P. Choudhury, T. Hart, B. Dolan, K. Toutanova, M. Mitchell, Microsummarization of online reviews: an experimental study, in *Proceedings of the AAAI Conference on Artificial Intelligence (AAAI)* (2016), pp. 3015–3021
12. J. McAuley, J. Leskovec, Hidden factors and hidden topics: understanding rating dimensions with review text, in *Proceedings of the ACM Conference on Recommender Systems (RecSys)* (ACM, New York, 2013), pp. 165–172
13. J. McAuley, R. Pandey, J. Leskovec, Inferring networks of substitutable and complementary products, in *Proceedings of the ACM International Conference on Knowledge Discovery and Data Mining (SIGKDD)* (ACM, New York, 2015), pp. 785–794
14. A. Mukherjee, A. Kumar, B. Liu, J. Wang, M. Hsu, M. Castellanos, R. Ghosh, Spotting opinion spammers using behavioral footprints, in *Proceedings of the ACM International Conference on Knowledge Discovery and Data Mining (SIGKDD)* (ACM, New York, 2013), pp. 632–640
15. C. Musat, Y. Liang, B. Faltings, Recommendation using textual opinions, in *Proceedings of the International Joint Conference on Artificial Intelligence (IJCAI)* (2013), pp. 2684–2690
16. M. Ott, Y. Choi, C. Cardie, J.T. Hancock, Finding deceptive opinion spam by any stretch of the imagination, in *Proceedings of the Annual Meeting of the Association for Computational Linguistics (ACL)* (Association for Computational Linguistics, Stroudsburg, 2011), pp. 309–319
17. J. Pasternack, D. Roth, Knowing what to believe (when you already know something), in *Proceedings of the International Conference on Computational Linguistics (COLING)* (Association for Computational Linguistics, Stroudsburg, 2010), pp. 877–885
18. K. Saleh, The importance of online customer reviews [Infographic] (2017), https://www.invespcro.com/blog/the-importance-of-online-customer-reviews-infographic/. Accessed 2017-06-29
19. H. Sun, A. Morales, X. Yan, Synthetic review spamming and defense, in *Proceedings of the ACM International Conference on Knowledge Discovery and Data Mining (SIGKDD)* (ACM, New York, 2013), pp. 1088–1096
20. Y. Tan, M. Zhang, Y. Liu, S. Ma, Rating-boosted latent topics: Understanding users and items with ratings and reviews, in *Proceedings of the International Joint Conference on Artificial Intelligence (IJCAI)* (2016), pp. 2640–2646
21. D. Tang, B. Qin, T. Liu, Y. Yang, User modeling with neural network for review rating prediction, in *Proceedings of the International Joint Conference on Artificial Intelligence (IJCAI)* (2015), pp. 1340–1346
22. I.E. Vermeulen, D. Seegers, Tried and tested: the impact of online hotel reviews on consumer consideration. Tour. Manag. **30**(1), 123–127 (2009)
23. Y. Yang, Y. Yan, M. Qiu, F.S. Bao, Semantic analysis and helpfulness prediction of text for online product reviews, in *Proceedings of the Annual Meeting of the Association for Computational Linguistic (ACL) and the International Joint Conference on Natural Language Processing* (2015), pp. 38–44
24. K. Yatani, M. Novati, A. Trusty, K. Truong, Analysis of adjective-noun word pair extraction methods for online review summarization, in *Proceedings of the International Joint Conference on Artificial Intelligence (IJCAI)* (2011)

25. X. Yin, J. Han, S.Y. Philip, Truth discovery with multiple conflicting information providers on the web. IEEE Trans. Knowl. Data Eng. **20**(6), 796–808 (2008)
26. W. Zhang, Q. Yuan, J. Han, J. Wang, Collaborative multi-level embedding learning from reviews for rating prediction, in *Proceedings of the International Joint Conference on Artificial Intelligence (IJCAI)* (2016)
27. X. Zheng, W. Ding, J. Xu, D. Chen, Personalized recommendation based on review topics. Serv. Orient. Comput. Appl. **8**(1), 15–31 (2014)

Erratum to: Ontology-Based Data Access with Datalog+/−

Gerardo I. Simari, Cristian Molinaro, Maria Vanina Martinez,
Thomas Lukasiewicz, and Livia Predoiu

Erratum to:
Chapter 1 in: G.I. Simari et al., *Ontology-Based*
Data Access Leveraging Subjective Reports,
SpringerBriefs in Computer Science,
https://doi.org/10.1007/978-3-319-65229-0_1

The author of the book provided the below additional reference and its respective citation for Chapter 1 after the book is published. This has now been updated in the respective chapter in the revised version of the book.

O. Tifrea-Marciuska, Personalised search for the social semantic web. D.Phil. Thesis, Department of Computer Science, University of Oxford (2016)

The updated online version of this chapter can be found at
https://doi.org/10.1007/978-3-319-65229-0_1

Printed in the United States
By Bookmasters